Collins

GCSE Revision

Maths

Higher

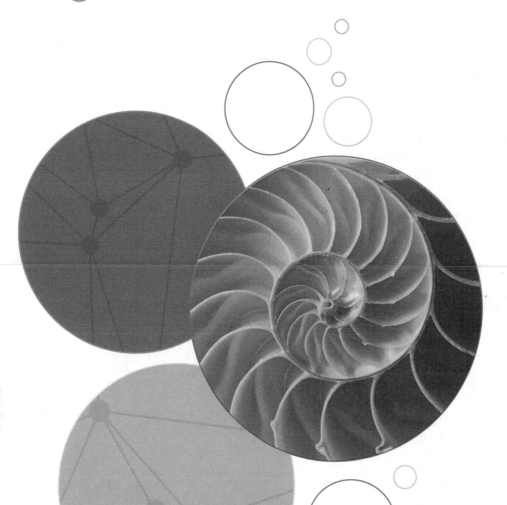

Maths

Higher

GCSE

Revision Guide

Linda Couchman
and Rebecca Evans

Contents

Contents

S Statistics   **P** Probability   **R** Ratio, Proportion and Rates of Change

Contents

N Number   **A** Algebra   **G** Geometry and Measures

Contents

Review Questions

Recap of KS3 Key Concepts

1 Write down two million and five as a number. [1]

2 Work out 15% of 300kg. [1]

3 Write 60 as the product of prime factors. [1]

4 Find **a)** the HCF and **b)** the LCM of 36 and 90. [2]

5 $6 + 4 \div 2 =$ _____ [1]

6 $(-2)^3 =$ _____ [1]

7 $(2^3)^2 =$ _____ [1]

8 Simplify: $5 + 2(m - 1)$ [1]

9 Simplify: $y^2 + y^2 + y^2$ [1]

10 **a)** Write the next two terms in the number sequence: 5, 8, 11, 14, __, __ [1]

    b) Is 140 a term in the sequence? Explain your answer. [2]

11 $0.3 \times 0.2 =$ _____ [1]

12 A regular polygon has an interior angle of 140°. How many sides does the polygon have? [1]

13 Work out $4.152 \div 1.2$ [1]

14 If $A = (2, 3)$ and $B = (7, 11)$, work out the coordinates of the midpoint of the line AB. [2]

15 $3 \times 3 \times 3 =$ _____ [1]

16 Lorna left her home at 11.00am and went for a 20km run. She arrived back home at 1.00pm.

Work out Lorna's average running speed. [1]

17 The cross-section of a prism has an area of 25cm². The length of the prism is 5cm.

Work out the volume of the prism. [1]

18 The probability of it raining is 0.6

Work out the probability of it **not** raining. [1]

19 Write $\frac{7}{8}$ as a decimal. [1]

20 Write down 0.78 as a fraction in its simplest form. [2]

21 Find a fraction that lies between $\frac{2}{3}$ and $\frac{3}{4}$. [2]

22 Pat threw three darts. The lowest score was 10. The range was 10. The mean was 15.

Work out the score for each dart. [3]

23 Write down the number that is one less than one million. [1]

24 Share £56 in the ratio 3 : 5 [2]

25 Put the following numbers in order, smallest first: 0.4, 0.04, 0.39, 0.394 [1]

26 Calculate a value for m if:

    a) $m - 5 = -1$ [1]

    b) $m \div 4 = 12$ [1]

    c) $2m + 13 = 16$ [1]

27 A square has a perimeter of 24cm. Work out its area. [1]

28 Work out $1616 \div 4$. [1]

29 Temi spends £7.25. She pays with a £20 note.

How much change does she receive? [1]

30 Simplify $10a - 6y + 3a - 4y$ [1]

31 Increase £60 by 35%. [1]

32 A circle has a radius of 10cm. Taking π as 3.14, work out the area of the circle. [2]

33 How many cubes of side length 2cm will fit inside a hollow cube of side length 4cm? [3]

34 Write down the square root of 225. [1]

35 $\frac{3}{4}$ of a number is 63. Write down the number. [2]

36 Work out the value of m if $2 - m = 5$ [1]

37 Give the coordinates of the point where the graph of $x + y = 5$ crosses the x-axis. [2]

38 The two short sides of a right-angled triangle are 5cm and 12cm.

Work out the length of the longest side. [1]

39 If $a = 3$ and $b = 4$, work out the value of $2a^2 + 3b$ [2]

40 Find the median of 1, 4, 6, 13, 21, 10, 3 and 7. [1]

Total Marks _____ / 57

Order and Value

You must be able to:

- Order and compare positive and negative numbers
- Carry out calculations using positive and negative integers and decimals
- Understand and use standard form notation.

Negative Numbers

Negative Numbers | Positive Numbers

−10 −9 −8 −7 −6 −5 −4 −3 −2 −1 0 1 2 3 4 5 6 7 8 9 10

- On a number line, positive numbers (+) are to the right of zero and negative numbers (−) are to the left of zero.
- Numbers on the left are smaller than numbers on the right.
- −4 is less than −2 and can be written −4 < −2
- −2 is greater than −4 and can be written −2 > −4
- A number line can help to visualise questions that involve negative numbers.

Key Point
A negative number is smaller than zero.

Put the following numbers in order, smallest to largest:

−10.2   −0.3   −$\frac{1}{5}$ −6.4

−10.2   −0.3   −0.2   −6.4   ← Change all the numbers into decimals.

−10.2   −6.4   −0.3   −0.2   ← Then rearrange by size.

Work out 2 − 7

finish   start

−6 −5 −4 −3 −2 −1 0 1 2 3

2 − 7 = −5

Work out −2 + (−2)

finish   start

−6 −5 −4 −3 −2 −1 0 1 2 3

−2 − 2 = −4

Key Point
−2 + (−2) and −2 − (+2) both mean −2 − 2 = −4

Work out −2 − (−2)   ← Two minus signs together become a plus.

−2 + 2 = 0

- When multiplying or dividing with two signs that are **different**, the answer is **negative**.
- When multiplying or dividing with two signs that are the **same**, the answer is **positive**.

$-3 \times 6 = -18$	$7 \times (-2) = -14$	$24 \div (-6) = -4$
$-5 \times (-7) = 35$	$-100 \div (-5) = 20$	

Calculating with Decimals

- When **multiplying**, remove the decimal points and multiply the whole numbers.
- The number of decimal places in the answer is the same as the **total** number of decimal places in the original calculation.

Work out 0.6×0.2

$6 \times 2 = 12$ ← Remove the decimal points.

$0.6 \times 0.2 = 0.12$ ← Two figures after the decimal points in the question, so two figures after the decimal point in the answer.

Work out 1.46×2.3

$146 \times 23 = 3358$

$1.46 \times 2.3 = 3.358$ ← Three figures after the decimal points on both sides of the equals sign.

- When **dividing**, change the problem so that you divide by a whole number.

Work out $27.632 \div 0.2$

$276.32 \div 2 = 138.16$ ← In order to divide by a whole number, multiply everything by 10.

Work out $36.14 \div 0.002$

$36\,140 \div 2 = 18\,070$ ← Multiply everything by 1000.

Standard Form

- **Standard form** is used to represent very small or very large numbers.
- A number in standard form is written in the form $A \times 10^n$, where $1 \leqslant A < 10$ and n is an integer.
- For numbers less than 1, n is negative.

Write the following numbers in standard form:

a) $723\,000$ $723\,000 = 7.23 \times 10^5$ ← $n = 5$ since the decimal point has to move 5 places to the right to go from 7.23 to 723 000

b) $0.000\,0063$ $0.000\,0063 = 6.3 \times 10^{-6}$ ← $n = -6$ since the decimal point has to move 6 places to the left to go from 6.3 to 0.000 0063

> **Quick Test**
>
> 1. Put the following numbers in order, smallest first:
> 0.0125 $\frac{1}{25}$ 1.25 -0.125
> 2. a) $\frac{-6}{-0.25}$ b) $-6 - (-7)$ c) $-\frac{1}{2} + 3 \times (-4)$ d) $(-1)^{20}$
> 3. Write the following numbers in standard form:
>   a) 0.00163 b) $16\,300\,000$

> **Key Words**
>
> negative number
> positive number
> standard form

Types of Number

You must be able to:

- Use the ideas of prime numbers, factors (divisors), multiples, highest common factor, lowest common multiple and prime factors
- Show prime factor decomposition
- Use systematic listing strategies, including the product rule for counting.

Types of Number

- **Multiples** are found in the 'times table' of the number, e.g.
  Multiples of 4 = {4, 8, 12, 16 …}
- A **factor** (divisor) is a number that will divide exactly into another number, e.g.
  Factors of 12 = {1, 2, 3, 4, 6, 12}
- A **prime number** has only two factors: itself and 1.
- **Square numbers** are the results of multiplying together two numbers that are the same. They are shown using a **power** of 2, e.g.

$1 \times 1 = 1$	$2 \times 2 = 4$	$3 \times 3 = 9$	$4 \times 4 = 16$
$1^2 = 1$	$2^2 = 4$	$3^2 = 9$	$4^2 = 16$

- A **square root** is the **inverse** (opposite) of a square, e.g.
  $6^2 = 36$ $\sqrt{36} = 6$
- **Cube numbers** are the results of multiplying together three numbers that are the same. They are shown using a power of 3, e.g.

$1 \times 1 \times 1 = 1$	$2 \times 2 \times 2 = 8$	$3 \times 3 \times 3 = 27$	$4 \times 4 \times 4 = 64$
$1^3 = 1$	$2^3 = 8$	$3^3 = 27$	$4^3 = 64$

- A **cube root** is the inverse of a cube, e.g.
  $5^3 = 125$ $\sqrt[3]{125} = 5$

> **Key Point**
>
> When listing factors, look for pairs of numbers, then you will not forget any.

> **Key Point**
>
> You are expected to know the squares and square roots up to $15 \times 15 = 225$ and that $\sqrt[3]{1} = 1$, $\sqrt[3]{8} = 2$, $\sqrt[3]{27} = 3$, $\sqrt[3]{64} = 4$, $\sqrt[3]{125} = 5$ and $\sqrt[3]{1000} = 10$

Prime Factors, LCM and HCF

- The **lowest common multiple (LCM)** of two numbers is the lowest **integer** that is a multiple of both numbers.

  Find the LCM of 12 and 18.

  Multiples of 12 = {12, 24, (36), 48 …}   ← Write out the multiples of 12 and 18 until you get a common value.

  Multiples of 18 = {18, (36), 54 …}

  36 is the smallest number that is in both lists, so 36 is the LCM.

- The **highest common factor (HCF)** of two numbers is the largest integer that will divide exactly into both numbers.

  Find the HCF of 45 and 60.

  Factors of 45 = {1, 3, 5, 9, (15), 45}   ← Write out the factors of 45 and 60 and look for the highest common value.

  Factors of 60 = {1, 2, 3, 4, 5, 6, 10, 12, (15), 20, 30, 60}

  15 is the largest number that is in both lists, so 15 is the HCF.

- Factors of a number that are also prime numbers are called **prime factors**.

Write 48 as a product of prime factors.

Method 1:
Prime Factor Decomposition

2	48
2	24
2	12
2	6
3	3
	1

Method 2:
A Factor Tree

> Keep splitting values into factors until the ends of all branches are prime numbers.

> Keep dividing by prime numbers until you get to 1.

$$48 = 2 \times 2 \times 2 \times 2 \times 3 = 2^4 \times 3$$

Choices and Outcomes

- The **product rule** (multiplication rule) for counting is used to find the total number of **combinations** or **permutations** possible for a given scenario.
- It states: if there are A ways of doing Task 1 and B ways of doing Task 2, then there are $A \times B$ ways of doing both tasks.

In a restaurant there are three different starters (X, Y, Z) and six different main meals (1, 2, 3, 4, 5, 6) to choose from.

How many possible combinations are there when choosing a starter and a main meal?

Combinations are: X1, X2, X3, X4, X5, X6, Y1, Y2, Y3, Y4, Y5, Y6, Z1, Z2, Z3, Z4, Z5, Z6 (18 combinations)

Using the product rule:
3 (starters) × 6 (mains) = 18 (combinations)

How many different ways can you rearrange the letters PENCIL?

$6 \times 5 \times 4 \times 3 \times 2 \times 1 = 720$ ways ← For the first letter, there is a choice of six, for the second letter, a choice of five, for the third letter, a choice of four, etc.

Key Point

A permutation is an arrangement of objects where the order is important. If order does not matter, it is a combination.

Key Words

multiple
factor
prime number
square number
power
square root
cube number
cube root
lowest common multiple (LCM)
integer
highest common factor (HCF)
prime factor
product rule
combination
permutation

Quick Test

1. Express 36 as a product of prime factors using any method.
2. Find **a)** the HCF and **b)** the LCM of 12 and 15.
3. Red buses leave the bus garage every 12 minutes. Blue buses leave the bus garage every 20 minutes. A red bus and a blue bus both leave the garage at 9am. At what time will a red bus and a blue bus next leave the garage together?

Basic Algebra

You must be able to:

- Use and understand algebraic notation and vocabulary
- Simplify and rearrange expressions, including those that involve algebraic fractions
- Expand and factorise expressions
- Solve linear equations.

Basic Algebra

- 'Like terms' are **terms** with the same **variable**, e.g. $3x$ and $4x$ are like terms because they both contain the variable x.
- To simplify an **expression** you must collect like terms.
- When moving a term from one side of an **equation** to the other, you must carry out the **inverse operation**.

Simplify $3x + 3y - 7x + y$

$4y - 4x$ ←

$3x - 7x = -4x$
$3y + y = 4y$

Simplify $9p^2 + 7p - qp + pq - p^2$ ←

$8p^2 + 7p$ ←

$qp = pq$

$9p^2 - p^2 = 8p^2$
$-qp + pq = 0$

- Substitution means replacing variables with numbers.

If $y = 4$ and $t = 6$, work out the value of $7y - 6t$

$7y - 6t = 7 \times 4 - 6 \times 6$
$\qquad = 28 - 36$
$\qquad = -8$

If $q = 5$, $r = 2$ and $z = -3$, work out the value of $rq + z^2$

$rq + z^2 = 2 \times 5 + (-3)^2$ ←

$\qquad = 10 + 9$
$\qquad = 19$

- To expand or multiply out brackets, every term in the bracket is multiplied by the term outside the bracket.

Multiply out $5p(p - 2)$

$5p^2 - 10p$ ←

Expand and simplify $4y(2y - 3) - 3y(y - 2)$

$8y^2 - 12y - 3y^2 + 6y$ ←

$= 5y^2 - 6y$

Key Point

When simplifying expressions, remember to:

- Use BIDMAS
- Show your working.

Key Point

An expression does not contain an = sign.

Use brackets as the minus sign is also squared.

Key Point

Always apply the rules:

$- \times - = +$
$+ \times + = +$
$- \times + = -$
$+ \times - = -$

$5p \times p = 5p^2$
$5p \times (-2) = -10p$

Note that $-3y \times -2 = +6y$

Factorisation

- **Factorisation** is the reverse of expanding brackets, i.e. you take out a common factor and put brackets into the expression.
- To factorise, you should look for common factors in every term.

Factorise $3x^2 - 6x$	Factorise $3p^3 - 2p^2 + 8p$
$3x(x - 2)$	$p(3p^2 - 2p + 8)$

> **Key Point**
>
> To factorise completely, always take out the highest common factor, e.g. 3 is the HCF of 3 and 6.

Linear Equations

- Equations can be used to represent real-life problems.
- The equation should be rearranged to solve the problem.

Remember $x^2 = x \times x$

Solve $5y - 4 = 3y + 10$

$$5y - 3y = 10 + 4$$
$$2y = 14$$
$$y = 7$$

Collect all the letter terms on one side.

> **Key Point**
>
> When moving a term from one side of an equation to the other, you must carry out the inverse operation.

Mary bought nine candles. She used a £3 gift voucher as part payment. The balance left to pay was £5.55

What was the cost of one candle (c)?

$$9c - 3 = 5.55$$
$$9c = 8.55$$
$$c = \frac{8.55}{9}$$
$$c = £0.95 \text{ or } 95p$$

Use the information given to set up an equation.

Solve to find the cost of one candle.

Algebraic Fractions

- When solving equations involving **fractions**, take extra care when rearranging and make sure you carry out the inverse operations in the correct order.

Solve $\frac{3x - 3}{4} = 7$	Solve $\frac{5s}{4} + 3 = 18$
$3x - 3 = 28$	$\frac{5s}{4} = 15$
$3x = 31$	$5s = 60$
$x = \frac{31}{3}$	$s = 12$

Subtract 3 from both sides.

Multiply both sides by 4.

Divide both sides by 5.

> **Key Words**
>
> term
> variable
> expression
> equation
> inverse operation
> factorisation
> fraction

> **Quick Test**
>
> 1. Simplify $2y - 7 + 4y + 2$
> 2. Work out the value of $3p^3 - 7q$, when $p = -4$ and $q = -3$
> 3. Expand the following expression: $3t(4t - 1)$
> 4. Factorise completely $4r^3 - 2r^2$
> 5. Solve the equation $5(t + 4) = 3(6 - t)$

Factorisation and Formulae

You must be able to:

- Expand products of two or more binomials
- Factorise a quadratic expression
- Understand and use formulae
- Rearrange and change the subject of a formula.

Binomial Expansion

- A **binomial** is an **expression** that contains two terms, e.g. $3y^2 + 12$, $4xy - x^2$ or $5x^3 + 3x^2$
- The product of binomials is obtained when two or more binomials are multiplied together, e.g. $(2r + 7)(3r - 6)$
- To **expand** (or multiply out) the brackets, every term in the first set of brackets must be multiplied by every term in each of the other sets of brackets.

Expand $(2y + 4)(3y - 2)$

×	$2y$	$+4$
$3y$	$6y^2$	$+12y$
-2	$-4y$	-8

$6y^2 + 12y - 4y - 8$
$= 6y^2 + 8y - 8$

Expand $(4z + 3)(2z - 1)(4 - z)$

×	$4z$	$+3$
$2z$	$8z^2$	$+6z$
-1	$-4z$	-3

Multiply out the first two brackets.

×	$8z^2$	$+2z$	-3
4	$32z^2$	$+8z$	-12
$-z$	$-8z^3$	$-2z^2$	$+3z$

$8z^2 + 6z - 4z - 3$
$= 8z^2 + 2z - 3$

$32z^2 + 8z - 12 - 8z^3 - 2z^2 + 3z$
$= -8z^3 + 30z^2 + 11z - 12$

Key Point

Make sure you multiply every term inside the bracket.

Key Point

Take care over + and – signs.

Remember $x^m \times x^n = x^{m+n}$

> Simplify by collecting like terms.

> Multiply the product of the first two brackets with the final bracket.

> Write in descending powers of z.

Quadratic Factorisation

- An expression that contains a squared term is called **quadratic**.
- Some quadratic expressions can be written as a product of two binomial expressions.
- When written in factorised form, the new expression is equivalent to the original quadratic.
- To factorise a quadratic, you can use a table as in the expansion examples above:
  1. Complete the table with as much information as you can.
  2. Work out the missing terms (it may take more than one attempt to find the correct pair of numbers).
  3. Write out the factorised expression.

Key Point

Check you have factorised correctly by expanding the brackets – your expressions should be equivalent.

- If the **coefficient** of x^2 is **not** 1, more care must be taken.

Factorise the expression
$x^2 + 4x + 3$

×	x	$+1$
x	x^2	$+x$
$+3$	$+3x$	$+3$

The missing terms need to have a product of +3 and a sum of +4, i.e. 1 and 3.

$(x + 1)(x + 3)$

Write the expression as a product: the **first row** gives you the **first bracket** and the first column gives you the **second bracket**.

Factorise the expression
$2y^2 - 3y - 20$

×	$2y$	$+5$
y	$2y^2$	$+5y$
-4	$-8y$	-20

$(2y + 5)(y - 4)$

+5 and −4 is the pair that gives −3y.

Note that the coefficient of y^2 is not 1.

Possible pairs of numbers are −20 and 1; 20 and −1; 10 and −2; −10 and 2; −4 and 5; −5 and 4.

Changing the Subject of a Formula

- A **formula** is a way of describing a rule or fact.
- A **formula** is written as an algebraic equation.
- The **subject** of a formula appears once on the left-hand side.
- To change the subject, a formula must be rearranged using **inverse operations**.

This formula can be used to change temperature in degrees Fahrenheit to temperature in degrees Celsius: $C = \frac{5}{9}(F - 32)$

In Iceland, the lowest recorded temperature on a certain day is −20°C. What is this temperature in degrees Fahrenheit?

$-20 = \frac{5}{9}(F - 32)$

$-180 = 5(F - 32)$

$-36 = F - 32$

$F = -4°F$

The formula must be rearranged to find the value of F.

The answer is −4 degrees Fahrenheit.

Make r the subject of the formula $P = 3(r - 1)$

$\frac{P}{3} = r - 1$

$r = \frac{P}{3} + 1$

The formula for calculating the area of a circle is $A = \pi r^2$
Make r the subject.

$\frac{A}{\pi} = r^2$

$r = \sqrt{\frac{A}{\pi}}$

π can be treated as a numerical term.

Only the positive root is needed as r is a length.

Collect all the terms containing q on one side.

Make q the subject of the formula $p + q = pq - 5$

$p + 5 = pq - q$

$p + 5 = q(p - 1)$

$q = \frac{p + 5}{p - 1}$

Take q out as a factor.

Quick Test

1. $T = 30w + 20$. Work out the value of w when $T = 290$
2. Expand and simplify $(y + 4)(y - 2)$
3. Factorise $2q^2 + 7q + 3$
4. Make y the subject of the formula $\frac{x + y}{3} = 2(y - 1)$

Ratio and Proportion

You must be able to:

- Apply ratio to real contexts and problems, including best buys
- Divide quantities into given ratios
- Use direct proportion
- Understand graphs that illustrate direct and inverse proportion.

Best Buys

- Working out the value of one part allows you to make like-for-like comparisons.

> Teabags are sold in two different sized packs.
>
> Pack A holds 80 teabags and costs £1.80
>
> Pack B holds 200 teabags and costs £3.80
>
> Which is the best buy?
>
> Pack A   180 ÷ 80 = 2.25p
>
> Pack B   380 ÷ 200 = 1.90p
>
> Pack B is the best buy (the cheapest price per teabag).

Key Point

If a ratio contains mixed units, change all values to the same unit before simplifying.

Work out the cost of one teabag in each pack.

Ratio

- You can **simplify** a **ratio**. This is like cancelling down a fraction.
- The unitary form of a ratio is $1 : n$

> Write down the ratio 5 : 9 in its unitary form.
>
> 5 : 9 = 1 : 1.8

Divide both sides by 5.

- To divide a quantity into a given ratio, you must first work out the value of one part.

> Sarah, John and James share £300 in the ratio 3 : 4 : 5
>
> How much money does each person receive?
>
> 3 + 4 + 5 = 12 parts in total, so £300 = 12 parts
>
> 300 ÷ 12 = £25
>
>   3 : 4 : 5 = (3 × £25) : (4 × £25) : (5 × £25)
>
> Sarah : John : James = £75 : £100 : £125

Divide by 12 to find the value of one part.

Always add up the values in your answer to check they equal the starting amount.

> An apple crumble recipe for four people uses 80g of plain flour.
>
> How much flour is needed for six people?
>
> 1 person uses 80 ÷ 4 = 20g
>
> 6 people use 6 × 20g = 120g

Divide by 4 to find the value of one part.

Direct and Inverse Proportion

- In a graph, **direct proportion** can be represented by a straight line that passes through the origin (0, 0).

This table gives information about the journey of a car during a half-hour time period.

Distance (miles)	0	6	12	18
Time (mins)	0	10	20	30

a) Plot the graph of distance against time.

b) Is the distance directly proportional to time? Give a reason for your answer.

c) At what speed was the car travelling in mph?

a)

b) Distance is directly proportional to time, because it produces a straight line graph that passes through the origin.

c) Speed = $\dfrac{\text{Distance}}{\text{Time}}$

$= \dfrac{18 \text{ miles}}{30 \text{ minutes}}$

$= \dfrac{36 \text{ miles}}{60 \text{ minutes}}$

$= 36\text{mph}$

This is the gradient.

- Quantities are in direct **proportion** if their ratio remains the same as they are increased or decreased.
- Direct proportion uses the symbol α.
- $y \, \alpha \, x$ means y is directly proportional to x or $y = kx$.
- k is called the **constant of proportionality**.
- Quantities can also be in **inverse proportion**. This is shown by $y \, \alpha \, \frac{1}{x}$ or $y = \frac{k}{x}$
- For example, for a journey of fixed distance (D), the time taken (T) is inversely proportional to the speed of travel (S), i.e. $T \, \alpha \, \frac{1}{s}$

Key Point

If quantities are in inverse proportion, as x increases, y decreases. The graph produced is a hyperbola (curve).

Quick Test

1. £180 is shared in the ratio 2 : 3
   What is the value of the largest share?
2. Simplify 3 hours : 45 minutes
3. a) Cat food comes in two different sized boxes:
   Box A has 12 sachets and costs £3.80
   Box B has 45 sachets and costs £14.00
   Which is the best buy? Explain your answer clearly.
   b) Tiggles the cat eats three sachets a day. How many of the best buy boxes have to be bought to feed Tiggles in June?

Key Words

simplify
ratio
direct proportion
proportion
constant of
    proportionality
inverse proportion

Variation and Compound Measures

You must be able to:

- Use compound units and solve problems involving compound measures
- Calculate compound interest
- Use iterative processes.

Compound Measures

- A **compound measure** is a measure that involves two or more other measures.
- **Speed** is a compound measure – it involves measures of distance and time.

$$\text{Speed } (S) = \frac{\text{Total Distance } (D)}{\text{Time Taken } (T)} \qquad D = S \times T \qquad T = \frac{D}{S}$$

A snail crawls a distance of 30cm in 3 minutes 20 seconds. Work out its speed in m/s.

30cm = 0.3m ⟵ Change 30cm into metres by dividing by 100 (100cm = 1m).

3 minutes 20 seconds = 200 seconds ⟵

Speed = Distance ÷ Time

Change the total time into seconds.

= 0.3 ÷ 200

= 0.0015m/s

- **Density** and **pressure** are compound measures.

$$\text{Density } (D) = \frac{\text{Mass } (M)}{\text{Volume } (V)} \qquad M = D \times V \qquad V = \frac{M}{D}$$

$$\text{Pressure } (P) = \frac{\text{Force } (F)}{\text{Area } (A)} \qquad F = P \times A \qquad A = \frac{F}{P}$$

Compound Interest and Repeated Percentage Change

- **Compound interest** is calculated based on the total amount of money invested, plus any interest previously earned.
- This formula can be used to calculate how money invested grows with time:

$$\text{Final Amount } (A) = \text{Original Amount} \times \left(1 + \frac{\text{Rate}}{100}\right)^{\text{time}}$$

- To calculate **depreciation**, the plus sign in the formula is changed to a minus sign.

Key Point

Units for speed include kilometres per hour (km/h) or metres per second (m/s).

Key Point

Always check that the units in the answer are the same as the units required in the question.

Key Point

Units for density include kg/m^3 or g/cm^3.

Pressure is expressed in newtons (N) per square metre (m^2). A force of 1N applied to $1m^2$ is called 1 pascal (Pa).

Key Point

Simple interest is when the interest is paid out, so each year starts with the same amount.

£3000 is invested at 2% compound interest per annum.
Calculate how much money there will be after four years.
Give your answer to the nearest pound.

$$A = 3000 \times \left(1 + \frac{2}{100}\right)^4$$

$$= 3000 \times (1.02)^4$$

$$= 3000 \times 1.0824$$

$$= £3247 \text{ (to the nearest £)}$$

Use the formula:
$$A = \text{Original Amount} \times \left(1 + \frac{\text{Rate}}{100}\right)^{\text{time}}$$

Key Point

Compound interest questions only appear on a calculator paper.

The value of a new car is £8000.
The car depreciates in value by 10% each year.
Work out the car's value after five years.

$$A = 8000 \times \left(1 - \frac{10}{100}\right)^5$$

$$= 8000 \times (0.9)^5$$

$$= 8000 \times 0.5905$$

$$= £4723.92$$

Use the formula:
$$A = \text{Original Amount} \times \left(1 - \frac{\text{Rate}}{100}\right)^{\text{time}}$$

Trial and Improvement

- You can use trial and improvement (repetition) to find unknown values in a formula.

£600 is invested at $r\%$ compound interest.
After five years the value of the money has increased to £695.56

Using trial and improvement, find the value of r.

$$600 \times \left(1 + \frac{r}{100}\right)^5 = £695.56$$

Try $r = 6$ $600 \times (1.06)^5 = £802.94$ Too big.

Try $r = 4$ $600 \times (1.04)^5 = £729.99$ Slightly too big.

Try $r = 3$ $600 \times (1.03)^5 = £695.56$ Correct, so $r = 3\%$.

$$r = 3\%$$

Quick Test

1. An aircraft travels 134 miles in 20 minutes. What is the aircraft's speed in mph?
2. Work out the compound interest on £1200 invested at 1.4% per annum for three years.
3. A new plasma television is worth £429. If you want to sell it two years later, it will be worth £267.74. Use trial and improvement to find the rate of depreciation.

Key Words

compound measure
speed
density
pressure
compound interest
depreciation

Angles and Shapes 1

You must be able to:

- Recognise relationships between angles
- Use the properties of angles to work out unknown angles
- Recognise different types of triangle
- Understand and use the properties of special types of quadrilaterals.

Angle Facts

- There are three types of angle:
  - **acute**: less than 90°
  - **obtuse**: between 90° and 180°
  - **reflex**: between 180° and 360°.
- Angles on a straight line add up to 180°.
- Angles around a point add up to 360°.
- Vertically opposite angles are equal.

Angles in Parallel Lines

- Parallel lines never meet. The lines are always the same distance apart.
- **Alternate** angles are equal.
- **Corresponding** angles are equal.
- Co-interior or **allied** angles add up to 180°.

Alternate Angles

Corresponding Angles

Allied Angles

$c + d = 180°$

Work out the sizes of angles a, b, c and d.
Give reasons for your answers.

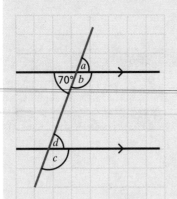

$a = 70°$ (vertically opposite angles are equal)

$b = 110°$ (angles on a straight line add up to 180°, so $b = 180° - 70°$)

$c = 110°$ (corresponding to b; corresponding angles are equal)

$d = 70°$ (corresponding to a; corresponding angles are equal)

Triangles

- Angles in a triangle add up to 180°.
- There are several types of triangle:
  - **equilateral**: three equal sides and three equal angles of 60°
  - **isosceles**: two equal sides and two equal angles (opposite the equal sides)
  - **scalene**: no sides or angles are equal
  - **right-angled**: one 90° angle.

> **Key Point**
>
> Examiners will **not** accept terms like 'Z angles' or 'F angles'. Always use correct terminology when giving reasons.

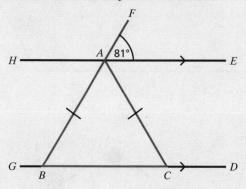

ABC is an isosceles triangle and *HE* is parallel to *GD*.

BAF is a straight line. Angle *FAE* = 81°

Calculate **a)** angle *ABC* and **b)** angle *ACB*.

Give reasons for your answers.

There are several different ways of solving this question.

a) Angle *HAB* = 81° (vertically opposite *FAE*), so angle *ABC* = 81° (alternate angle to *HAB*)

b) Angle *ACB* = 81° (angle *ABC* = angle *ACB*; base angles of an isosceles triangle are equal.)

Special Quadrilaterals

- The interior angles in a quadrilateral add up to 360°.
- You need to know the properties of these special quadrilaterals:

	Sides	Angles	Lines of Symmetry	Rotational Symmetry	Diagonals
parallelogram	opposite sides are equal and parallel	diagonally opposite angles are equal	none	order 2	diagonals bisect each other
rhombus	all sides are equal and opposite sides are parallel	opposite angles are equal	two	order 2	diagonals bisect each other at 90°
kite	two pairs of adjacent sides are equal	one pair of opposite angles is equal	one	none	diagonals cross at 90°
trapezium	one pair of opposite sides is parallel		none (an isosceles trapezium has one)	none	

Key Words

acute	isosceles
obtuse	scalene
reflex	right-angled
alternate	parallelogram
corresponding	rhombus
allied	kite
equilateral	trapezium

Angles and Shapes 2

You must be able to:

- Work out angles in a polygon
- Answer questions on regular polygons
- Understand and use bearings.

Angles in a Polygon

- A **polygon** is a closed shape with at least three straight sides.
- **Regular** polygons are shapes where all the sides and angles are equal.
- **Irregular** polygons are shapes where some or all of the sides and angles are different.
- For all polygons:
  - at any **vertex** (corner): **interior** angle + **exterior** angle = 180°
  - sum of all exterior angles = 360°
- To work out the sum of the interior angles in a polygon, you can split it into triangles from one vertex.
- For example, a pentagon is divided into three triangles, so the sum of the interior angles is 3 × 180° = 540°
- The sum of the interior angles for any polygon can be calculated using the formula:

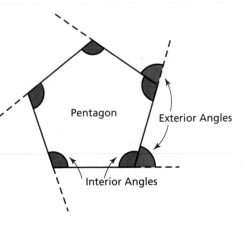

Pentagon   Exterior Angles

Interior Angles

> **LEARN**
>
> Sum of Interior Angles = $(n - 2) \times 180°$
> where n = number of sides

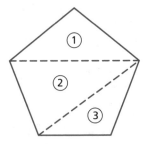

Work out the sum of the interior angles of a decagon (10 sides).

Sum = (10 – 2) × 180°   ⟵   Use the formula: Sum = $(n - 2) \times 180°$
    = 8 × 180°
    = 1440°

Regular Polygons

- In regular polygons:

> **LEARN**
>
> Number of Sides (n) × Exterior Angle = 360°
> So, Exterior Angle = 360° ÷ n

Work out the size of the interior angles in a regular hexagon (6 sides).

Exterior angle	= 360° ÷ 6 = 60°
Interior angle + 60°	= 180°
Interior angle	= 180° – 60°
	= 120°

Use the formula:
Exterior Angle = 360° ÷ n

Interior Angle + Exterior Angle = 180°

A regular polygon has an interior angle of 156°.

Work out the number of sides that the polygon has.

Exterior angle = 180° − interior angle
= 180° − 156°
= 24°
Number of sides = 360° ÷ 24°
= 15

Scale Drawings and Bearings

- **Bearings** are always measured in a clockwise direction from north (000°).
- Bearings always have three figures.

A ship sails from Mevagissey on a bearing of 130° for 22km.

a) Draw an accurate diagram to show this information and state the scale you have used.

b) What bearing would take the ship back to the harbour?

1cm : 10km

New bearing to return to harbour = 310° ◄───

Key Point

Always place the 0 to 180 line of the protractor onto the north-south line.

Measure with a protractor.

Quick Test

1. For a regular icosagon (20 sides), work out **a)** the sum of the interior angles and **b)** the size of one interior angle.
2. A regular polygon has an interior angle of 150°. How many sides does the polygon have?
3. Two yachts leave port at the same time.
   Yacht A sails on a bearing of 040° for 35km.
   Yacht B sails on a bearing of 120° for 60km.
   Using a scale of 1cm : 10km, draw the route taken by both yachts.
   What is the bearing of yacht B from yacht A?

Key Words

polygon
regular
irregular
vertex
interior
exterior
bearing

Fractions

You must be able to:

- Add, subtract, multiply and divide fractions
- Change terminating and recurring decimals into corresponding fractions and vice versa
- Express one quantity as a fraction of another.

Adding, Subtracting and Calculating with Fractions

- When the **numerator** and **denominator** of a fraction are both multiplied or divided by the same number, the result is an **equivalent** fraction.
- To add or subtract fractions, you must first convert them to equivalent fractions with the same denominator.

$$5\frac{3}{5} + 1\frac{2}{7}$$

$$= 6\frac{21}{35} + \frac{10}{35}$$

$$= 6\frac{31}{35}$$

$$4\frac{1}{4} - 2\frac{3}{5}$$

$$= \frac{17}{4} - \frac{13}{5}$$

$$= \frac{85}{20} - \frac{52}{20}$$

$$= \frac{33}{20} = 1\frac{13}{20}$$

> **Key Point**
>
> The top number in a fraction is called the **numerator**.
>
> The bottom number in a fraction is called the **denominator**.

Add whole numbers and then change fractions so that they have the same denominator.

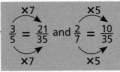

$$\frac{3}{5} = \frac{21}{35} \text{ and } \frac{2}{7} = \frac{10}{35}$$

Convert mixed numbers into improper fractions and then change fractions so that they have the same denominator.

$$\frac{17}{4} = \frac{85}{20} \text{ and } \frac{13}{5} = \frac{52}{20}$$

Multiplying and Dividing Fractions

- When multiplying fractions:
  - Convert mixed numbers to improper fractions
  - Cancel diagonally (or vertically) only
  - Then multiply the numerators together and the denominators together.
- When dividing fractions:
  - Invert the second fraction
  - Change the ÷ sign to a × sign
  - Then multiply.

> **Key Point**
>
> A **mixed number** contains a whole number and a fraction.
>
> An **improper fraction** has a numerator larger than the denominator.

Cancel down and then multiply. If the answer is an improper fraction, convert it back to a mixed number.

Rational Numbers, Reciprocals, and Terminating and Recurring Decimals

- **Rational numbers** are numbers that **can be written exactly** as a fraction or decimal, e.g. $\frac{1}{4} = 0.25$
- π is an **irrational number** as it would continue forever as a decimal 3.141592 …
- The **reciprocal** of a number (n) is 1 divided by the number, i.e. $\frac{1}{n}$
- A **terminating decimal** is a decimal with a finite number of digits (it ends), e.g. 0.75, 0.36
- A **recurring** decimal has a digit or group of digits that repeat indefinitely. It is shown by putting a dot over the recurring digit(s), e.g.
  $0.1\dot{6} = 0.166666…$ $0.\dot{3}\dot{7} = 0.37373737…$

Change $\frac{4}{9}$ into a recurring decimal.

$\frac{4}{9} = \frac{4.0000}{9}$ ← Divide 9 into 4.0000…

$= 0.4444…$

$= 0.\dot{4}$

Change $0.4\dot{3}$ into a fraction.

Let $d = 0.4\dot{3}$

$10d = 4.\dot{3}$ ← Multiply by 10.

$100d = 43.\dot{3}$ ← Multiply by 10 again.

$100d - 10d = 43.\dot{3} - 4.\dot{3}$ ← Subtract.

$90d = 39$

$d = \frac{39}{90} = \frac{13}{30}$

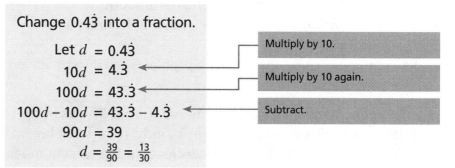

One Quantity as a Fraction of Another

Express 20 minutes as a fraction of 3 hours 20 minutes.

3 hours 20 minutes = 200 minutes ← Convert to minutes.

20 minutes as a fraction of 200 minutes $= \frac{20}{200} = \frac{1}{10}$ ← Cancel down.

A school has 712 students.
$\frac{5}{8}$ of the students travel to school by bus and the rest walk.
How many students walk to school?

Method 1

Work out $\frac{1}{8}$ of 712 and then multiply by 5 to get $\frac{5}{8}$

$712 \div 8 = 89$

$89 \times 5 = 445$ (number of students who take the bus to school)

Number who walk = 712 − 445
$\qquad\qquad\qquad = 267$ students

Method 2

$1 - \frac{5}{8} = \frac{3}{8}$

$\frac{3}{8} \times 712 = 267$ students

Start by working out how many students walk to school as a fraction.

Quick Test

1. Which is larger: $\frac{3}{4} + \frac{1}{5}$ or $\frac{3}{4} \div \frac{1}{5}$?
   You must show your working.
2. a) Change $\frac{1}{9}$ to a decimal.
   b) Is the resulting decimal terminating or recurring?
3. Express 15cm as a fraction of 4m in its simplest form.

Key Words

numerator
denominator
equivalent
rational number
irrational number
reciprocal
terminating decimal
recurring

Percentages

You must be able to:

- Increase and decrease quantities by a percentage
- Express one quantity as a percentage of another
- Solve percentage problems using a multiplier
- Understand and use reverse percentages.

Increasing and Decreasing Quantities by a Percentage

- When **increasing** a quantity by a **percentage**, work out the increase and then **add** it to the original amount.
- When **decreasing** a quantity, work out the decrease and then **subtract**.
- Percentage increase and decrease problems can also be solved using a **multiplier**.

> **Key Point**
>
> Percentages to learn:
> 100% = 1 (the whole)
> $20\% = \frac{1}{5}$
> $33.3\% = \frac{1}{3}$
> $66.7\% = \frac{2}{3}$
> $12.5\% = \frac{1}{8}$

Last year, Dandelion Place had 14 240 visitors. This year, the attraction will be closed for three months for refurbishment. This means there will be a 15% decrease in the total number of visitors for the year.

How many visitors are expected this year?

Method 1:

$\frac{15}{100} \times 14\,240 = 2136$

Number of visitors
= 14 240 – 2136
= 12 104

Method 2:

10% of 14 240 = 1424
5% of 14 240 = 712
15% of 14 240 = 2136
Therefore, the number of
visitors = 14 240 – 2136
= 12 104

| 10% ÷ 2 = 5% |

| 10% + 5% = 15% |

| Alternatively, using a multiplier: 15% decrease means there will be 85% of last year's visitors, so 0.85 × 14 240 = 12 104 |

Molly bought a new tumble dryer.
She visited three shops to compare prices:

A. Sean's Electricals
Normal price: £250, Sale price: 5% off

B. Mandeep's Deals
Tumble dryer £200, plus tax at 17.5%

C. Chet's Cut Price Store
Normal price: £280, Sale price: $\frac{1}{7}$ off

a) Which shop offered the best deal?
b) How much did it sell the tumble dryer for?

  A. Sean's Electricals:
  $\frac{5}{100} \times £250 = £12.50$
  Tumble dryer costs £250 – £12.50 = £237.50

B. Mandeep's Deals:
$\frac{17.5}{100} \times £200 = £35$ ←
Tumble dryer costs £200 + £35 = £235.00

C. Chet's Cut Price Store:
$\frac{1}{7} \times £280 = £40$
Tumble dryer costs £280 − £40 = £240.00

a) Mandeep's Deals b) £235

> If working without a calculator, remember:
> 17.5% = 10% + 5% + 2.5%

Sanjeev's salary is £32 000 per year. His salary is increased by 6%. Work out his new salary.

100% + 6% = 106%
    106% = 1.06 ←
1.06 × £32 000 = £33 920

> This is the multiplier.

Expressing One Quantity as a Percentage of Another

- Express the relationship between the two quantities as a fraction and then multiply by 100 to convert to a percentage.

Key Point

Make sure all quantities are in the same units first.

In two separate maths tests, Vinay got 18 out of 30 and Derek got 24 out of 45.

Who got the greater percentage in their maths test?

Vinay: $\frac{18}{30} \times 100 = 60\%$

Derek: $\frac{24}{45} \times 100 = 53.33\%$

Vinay got the greater percentage.

Work out 16 minutes as a percentage of 4 hours. Give your answer to 3 decimal places.

$\frac{16}{240}$ ←

$\frac{16}{240} \times 100 = 6.667\%$
(to 3 d.p.)

> 4 hours = 4 × 60 = 240 minutes

Reverse Percentages

- Reverse percentages involve working backwards from the final amount to find the original amount.

In a Thai restaurant, there were $27\frac{1}{2}$ dumplings left on a plate after 45% had been eaten. How many dumplings were on the plate at the start of the meal?

100% − 45% = 55% ←
     1% = 27.5 ÷ 55
  100% = 100 × 27.5 ÷ 55 = 50 dumplings

> $27\frac{1}{2}$ dumplings is 55% of the original quantity.

Quick Test

1. This year Curt grew 220 carrots. This is 20% fewer than last year. How many carrots did Curt grow last year?
2. Increase £60 by 17.5%.
3. Express 18cm as a percentage of 12m.

Key Words

percentage
multiplier

Probability 1

You must be able to:

- Know and use words associated with probability
- Construct and use a probability scale
- Understand what mutually exclusive events are
- Calculate probabilities using experimental data
- Categorise data into sets and subsets using tables and other diagrams.

Theoretical and Experimental Probability

Unlikely   Likely

Impossible   Evens   Certain
(0)   (0.5)   (1)

- The **probability** of an outcome occurring can be expressed using words or using a numerical scale from 0 to 1.
- **Relative frequencies** are probabilities based on experiments.
- **Random** means each possible outcome is equally likely.
- An event would be described as **biased** when outcomes are **not** equally likely.
- A sample space represents all possible outcomes from an event. This can be shown as a list or a diagram.

Key Point

Probabilities can be given as fractions, decimals or percentages.

Mutually Exclusive and Exhaustive Outcomes

- **Mutually exclusive** outcomes **cannot** happen at the same time.
- When two events are mutually exclusive P(A or B) = P(A) + P(B)
- A list of **exhaustive** events contains all possible outcomes.

Key Point

Probability is the chance of an event occurring. Probabilities can be based on theory or the results of an experiment. The sum of the probabilities of all possible outcomes is 1.

A spinner has three possible outcomes: A, B or C.
The probability that the spinner will land on A is 0.5
The probability that it will land on C is 0.3

a) Work out the probability that the spinner will land on B.

    0.5 + 0.3 = 0.8
    P(B) = 1 − 0.8 = 0.2

b) Work out the probability that the spinner lands on A or B.

    P(A or B) = 0.5 + 0.2 = 0.7

This is an exhaustive list.

The events are mutually exclusive.

Expectation

- When you know the probability of an outcome, you can predict how many times you would expect that outcome to occur in a certain number of trials. This is called **expectation**.

Key Point

P(A′) is the probability of A **not** occurring.

P(A′) = 1 − P(A)

Kelly rolls a six-sided dice 120 times and records her results.

Score	1	2	3	4	5	6
Frequency	19	29	14	18	20	20

a) Kelly throws the dice again. Estimate the probability that she will throw a 2.

Estimate of probability of a 2 is $\frac{29}{120}$

b) Is there enough evidence to suggest the dice is biased? Explain your answer.

If the dice is fair, you would expect Kelly to have rolled 20 of each number. This did not happen, so the dice may be biased. However, 120 is not a big sample. Kelly needs to roll the dice many more times. As she increases her sample size, her experiment will become more reliable and the results may come closer to the value of the theoretical probability.

Probability Diagrams

Julie thinks that in her class you are more likely to wear glasses if you are a boy. Based on the data below, is she correct?

	Boys	Girls	Total
Glasses	8	6	14
No Glasses	12	9	21
Total	20	15	35

P(of a boy wearing glasses) = $\frac{8}{20} = \frac{2}{5}$

P(of a girl wearing glasses) = $\frac{6}{15} = \frac{2}{5}$ ← Cancel down fractions.

Julie is incorrect as the probabilities are equal.

Revise

Students in a school study either French, Spanish, both or neither. There are 250 students. 75 students study both French and Spanish, 225 study French and 90 study Spanish.

Work out how many students do **not** study French or Spanish.

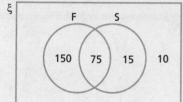

250 − 150 − 75 − 15 = 10

10 students do not study French or Spanish.

The crossover of the circles represents students who study both subjects.

Quick Test

1. A spinner has four faces labelled 1, 2, 3 and 4. The probability that it will land on a 2 or 4 is given in the table.

Number	1	2	3	4
Probability	x	0.3	x	0.5

   a) Work out the value of x.
   b) If you spin the spinner 150 times, how many times would you expect it to land on a 2?

Probability 2

You must be able to:

- Calculate the probability of independent and dependent combined events
- Calculate probabilities using a sample space diagram
- Calculate probabilities using a tree diagram
- Calculate and interpret conditional probabilities.

Addition Rules for Outcomes of Events

- When two outcomes are mutually exclusive, you can work out the probability of either outcome occurring by adding up the separate probabilities (see p.28).

Combined Events

- When calculating probabilities for combined events, you can draw a **sample space diagram**.

Two fair six-sided dice are thrown and their scores are added together.

a) Construct a sample space diagram to show all the possible outcomes.

		Dice 1				
+	1	2	3	4	5	6
1	2	3	4	5	6	7
2	3	4	5	6	7	8
3	4	5	6	7	8	9
4	5	6	7	8	9	10
5	6	7	8	9	10	11
6	7	8	9	10	11	12

(Dice 2 labels the left side)

b) Using your sample space diagram, work out P(8).

The sample space diagram shows there are $6 \times 6 = 36$ possible outcomes.

8 appears five times in the diagram out of 36 possible outcomes, therefore

$P(8) = \frac{5}{36}$

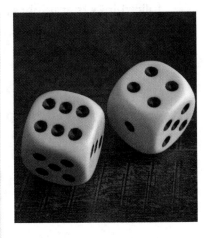

Tree Diagrams

- When dealing with successive events, you can use a **tree diagram**.

Two fair coins are tossed one after the other.

Write down the probability of each possible outcome.

0.5 — H
  0.5 — H   P(HH) = 0.5 × 0.5 = 0.25
  0.5 — T   P(HT) = 0.5 × 0.5 = 0.25
0.5 — T
  0.5 — H   P(TH) = 0.5 × 0.5 = 0.25
  0.5 — T   P(TT) = 0.5 × 0.5 = 0.25

Each coin can give two possible outcomes: a Head (H) or a Tail (T).

> **Key Point**
>
> Always multiply along the branches of a probability tree.

Probability of two Heads = P(HH) = 0.25
Probability of two Tails = P(TT) = 0.25
Probability of a Head and a Tail = P(HT) + P(TH)
$$= 0.25 + 0.25$$
$$= 0.5$$

0.25 + 0.25 + 0.5 = 1

Independent Events

- If A and B are **independent** events, one event does **not** depend on the other.
- If two events A and B are independent then P(A and B) = P(A) × P(B)

A fair two-sided coin is tossed and a fair six-sided dice is rolled.

What is the probability of getting a Head and rolling a 6?

$P(H) \times P(6) = \frac{1}{2} \times \frac{1}{6}$
$= \frac{1}{12}$

The two events are independent, therefore P(A) × P(B)

Conditional Probability

- If the probability of an event occurring is dependent on the outcome of the previous event, it is called **conditional** probability.

Philippa has a box of chocolates that contains 7 milk chocolates, 4 dark chocolates and 3 white chocolates.

If one of the milk chocolates is taken, only 13 chocolates remain and 6 of them will be milk.

She takes two chocolates from the box without looking. What is the probability that both chocolates are the same?

Probability that both are the same = P(MM) + P(DD) + P(WW)

$\left(\frac{7}{14} \times \frac{6}{13}\right) + \left(\frac{4}{14} \times \frac{3}{13}\right) + \left(\frac{3}{14} \times \frac{2}{13}\right) = \frac{30}{91}$

> **Quick Test**
>
> 1. A fair three-sided spinner has one red section, one green section and one blue section all of equal size. The spinner is spun and a fair six-sided dice is thrown. What is the probability of getting a 5 on the dice with a red on the spinner?
> 2. A bag contains three red balls, four blue balls and two yellow balls. A ball is drawn from the bag and **not** replaced. A second ball is then drawn from the bag. What is the probability that both balls are the same colour?

Key Words

sample space diagram
tree diagram
independent
conditional

Practice Questions

Order and Value

1. Write the following numbers in order, smallest to largest: 🖩

   −3   −3.38 $-\frac{1}{8}$ 3.3   [2]

2. Which is larger, −4 × −4 × −4 **or** $(2^3 − 3^3) − (−7)^2$? Show your working. 🖩   [3]

3. The average distance from the Moon to the Earth is 384 000km.

   a) Write this distance in standard form. 🖩   [2]

   b) A spaceship is travelling from the Earth to the Moon.
      It has travelled $1.96 × 10^4$km.

      How many kilometres are left to travel?   [2]

   > **Total Marks** _____ / 9

Types of Number

1. What is the largest multiple of 4 and 7 that is smaller than 105? 🖩   [1]

2. a) What is the lowest common multiple (LCM) of 12 and 15? 🖩   [1]

   b) What is the highest common factor (HCF) of 12 and 15?   [1]

3. Which is greater, the fifth square number or the third cube number?   [2]

4. 256 can be expressed as 2^n.

   What is the value of n?   [2]

5. Is 2^3 equal to 3^2?
   Give a reason for your answer. 🖩   [2]

6. Sausages come in packs of six. Bread rolls come in packs of four.

   What is the least number of packs of sausages and packs of bread rolls that Linda must buy,
   so that there is a roll for every sausage without any left over? 🖩   [2]

   > **Total Marks** _____ / 11

Basic Algebra & Factorisation and Formulae

1 Simplify $5x - 2y + 4x + 6y$ [2]

2 Simplify $5 - 3z + y - 5z + 7 + 3y$ [2]

3 Simplify $3x^2 + 3x + x^2 + 4 - x$ [2]

4 Work out the value of $3z^2 - 2q + 5$ when $z = -2$ and $q = -3$ [1]

5 Solve $4(2b - 3) = 2$ [2]

6 Solve $3(p + 2) = 2(p + 3)$ [2]

7 Solve $\frac{5}{2}x - \frac{1}{3} = \frac{2}{3}x + \frac{1}{2}$ [3]

8 Expand $6(x - 5y + 6)$ [2]

9 Expand $6p - 4(q - 3)$ [2]

10 Factorise completely $4xyz - 4xz$ [2]

11 Expand and simplify $(w + 4)(w + 1)(w - 4)$ [3]

12 Factorise $x^2 + 3x + 2$ [2]

13 Write $3(2x - 5) + 4(x + 3) - 4x$ in the form $a(bx + c)$, where a, b and c are integers. [3]

14 Rearrange the formula to make y the subject: $x = \dfrac{1 - 2y}{3 + 4y}$ [3]

15 The formula for the area of a trapezium is $A = \frac{1}{2}(a + b)h$

    a) Rearrange the formula to make h the subject. [2]

    b) The area of a trapezium is 24cm². Work out the value of h if $a = 5$cm and $b = 7$cm. [2]

16 Factorise $x^2 + 5x + 6$ [2]

17 $V = \sqrt{u^2 - 10p}$

    a) Work out the value of V when $u = 10$ and $p = 5$. [2]

    b) Rearrange the formula to make u the subject. [2]

Total Marks _____ / 41

Practice Questions

Ratio and Proportion

1 Simplify 5g : 10kg 🖩 [1]

2 The angles in a triangle are in the ratio 2 : 3 : 4

What is the size of the largest angle? 🖩 [2]

3 Six sticks of celery are needed to make celery soup for four people.

How many sticks of celery would be needed to make soup for 14 people? 🖩 [2]

4 It took six people four days to build a wall.

a) Working at the same rate, how long would it have taken eight people to build the wall? [2]

b) Working at the same rate, how many people would have been needed if the wall had to be completed in two days? [1]

Total Marks _____ / 8

Variation and Compound Measures

1 A bar of lead has a volume of 400cm³ and a mass of 4.56kg.

Work out the density of the bar of lead in g/cm³. [2]

2 A rabbit runs 200 metres in 22 seconds.

What is the rabbit's average speed in m/s?
Give your answer to 2 decimal places. [2]

3 Khalid left his home at 10am and went for a 15km run.
He arrived back home at 1pm.

What was his average speed in km/h? [2]

4 Work out the compound interest earned on £4000 invested at 4% for four years. [3]

Total Marks _____ / 9

Angles and Shapes 1 & 2

1 Work out the size of angles *j*, *k*, *l* and *m*, giving a reason for each answer.

[4]

2 *ABCD* is a parallelogram.
AB is parallel to *CD* and *AD* is parallel to *BC*. Angle *BAD* = 110°

Work out:

a) Angle *DCB*. [1]

b) Angle *ABC*. [1]

3 The angles in a quadrilateral are *x*, 2.5*x*, 3*x* and 2.5*x* degrees.

Calculate the size of the largest angle. [2]

4 Work out the interior angle of a regular decagon. [2]

5 A and B are two points.

If the bearing of B from A is 036°, what is the bearing of A from B? [1]

6 A regular polygon has an exterior angle of 45°.

a) Work out how many sides the polygon has. [1]

b) What is the name of the polygon? [1]

Total Marks _____ / 13

Fractions

1 A teacher took 32 books home to mark. She marked $\frac{1}{8}$ of them.

How many books does she still have to mark? [2]

2 On a farm, $\frac{1}{3}$ of the livestock is cows, $\frac{1}{6}$ is sheep, $\frac{1}{4}$ is chickens and the remainder is horses.

Express how many horses there are on the farm as a fraction. [2]

3 Write these fractions in ascending order:

$\frac{2}{3}$ $\qquad$ $\frac{4}{12}$ $\qquad$ $\frac{1}{6}$ $\qquad$ $\frac{12}{24}$ [2]

4 **a)** Work out the area of a rectangle measuring $\frac{4}{5}$ of a metre by $\frac{2}{3}$ of a metre. [1]

b) What is the rectangle's perimeter? Give your answer as an improper fraction. [2]

5 Change $0.1\dot{8}$ to a fraction. [2]

Total Marks _____ / 11

Percentages

1 Zahra earns £24 000 per year.
The first £6000 is tax free. She pays 22% tax on her remaining salary.

How much tax does she pay? [2]

2 Express 46 seconds of 2 hours as a percentage. [1]

3 Whizzy Garage sells cars. It offers a discount of 20% off the retail price for cash purchases.
Pat pays £4800 cash for a car.

Calculate the retail price of the car before the cash discount. [2]

Total Marks _____ / 5

Probability 1 & 2

1 Two fair six-sided dice are thrown and the product of their scores is calculated.

a) Draw a sample space diagram to represent all the possible outcomes. [2]

b) What is the probability that the product of the scores is greater than 12? [1]

c) What is the probability that the product of the scores is a square number? [1]

2 There are 20 pens in a box: ten black, six blue and four red.
A pen is taken at random from the box.

a) What is the probability that the pen is black? [1]

b) What is the probability that the pen is **not** black? [1]

3 A three-sided spinner has sides labelled 1, 2 and 3.
The spinner is biased.

Score	1	2	3
Frequency	98	305	97

Thomas spins the spinner 500 times and records the outcomes in a table.

a) Thomas spins the spinner again. Estimate the probability that it lands on a 2. [1]

b) He takes another 100 spins. Estimate how many times it will land on a 3. [2]

4 The probability that it will rain (R) on Wednesday is 0.65
If it does rain on Wednesday, the probability that it
will **not** rain (R') on Thursday is 0.37
If it does **not** rain on Wednesday, the probability it will
rain on Thursday is 0.55

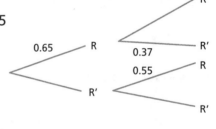

a) Complete the probability tree diagram. [3]

b) Calculate the probability that it rains on one of the days. [3]

5 Bhavna conducts a survey about the pets owned by her classmates.
There are 35 students in her class. Her results are: 25 students own cats,
15 students own dogs and 10 students own both.

a) Draw a Venn diagram to represent the results. [3]

b) How many students do **not** own a cat or a dog? [1]

Total Marks .... / 19

Number Patterns and Sequences

You must be able to:

- Work out missing terms in sequences using term-to-term rules and position-to-term rules
- Recognise and use arithmetic and geometric sequences
- Work out the rule for a given pattern.

Patterns in Number

- A **sequence** is a series of shapes or numbers that follow a particular pattern or rule.
- A **term-to-term rule** links the next term in the sequence to the previous one.
- A **position-to-term rule**, also called the *n*th term, can be used to work out any term in the sequence.
- A **recursive** relationship is used to define further terms in a sequence, when one or more initial terms are given.

> Write down the next two terms in the following sequence.
>
> 7, 11, 15, 19, __, __
>
> The term-to-term rule is +4, so the next two terms are 23 and 27.
>
> This sequence can be expressed by the recursive relationship
> $U_{n+1} = U_n + 4$, $U_1 = 7$
>
> This states that the first term is 7.
> Therefore, U_5 (the fifth term) = $U_4 + 4 = 19 + 4 = 23$

General Rules from Given Patterns

> Here is a sequence of patterns made from matchsticks:
>
>
>
> pattern 1   pattern 2   pattern 3
>
> **a)** Draw the next pattern in the sequence.
>
>
>
> **b)** Write down the sequence of numbers that represents the total number of matchsticks used in each pattern and work out an expression for the position-to-term rule.

The expression for the position-to-term rule is $3n + 1$

> | Pattern No. | 1 | 2 | 3 | 4 | n |
> |---|---|---|---|---|---|
> | No. of Matchsticks | 4 | 7 | 10 | 13 | $3n + 1$ |

Key Point

For the first term in any sequence, $n = 1$

The nth term of a number sequence is $5n + 2$.

Write down the first five terms of the sequence.

$(5 \times 1) + 2 = 7$ ←

$(5 \times 2) + 2 = 12$ ←

$7, 12, 17, 22, 27$ ←

To work out the first term, substitute $n = 1$ into the expression.

To work out the second term, substitute $n = 2$ into the expression.

Continue until $n = 5$ to produce the first five terms of the sequence.

Number Sequences

- In an **arithmetic sequence** there is a common difference between consecutive terms, e.g.

  $5, 8, 11, 14, 17 \ldots$ ←

  The terms have a common difference of $+3$.

- In a **geometric sequence** each term is found by multiplying the previous term by a constant, e.g.

  $20, 10, 5, 2.5, 1.25 \ldots$ ←

  The constant (or ratio) is 0.5

The nth term of an arithmetic sequence is $4n - 1$.

a) Write down the term-to-term rule.

The sequence of numbers is $3, 7, 11, 15, 19 \ldots$ ←

Work out the first five terms.

The term-to-term rule is $+4$.

b) Marnie thinks that 50 is a number in this sequence. Is Marnie correct? Explain your answer.

$4n - 1 = 50$, $n = 12.75$

n is not a whole number, so 50 is **not** in this sequence. Marnie is wrong.

Here is a geometric sequence: $4, 6, 9, \_\_, 20.25$

What is the missing term?

$\dfrac{6}{4} = \dfrac{9}{6} = 1.5$ ←

Divide at least two given terms by the previous term to work out the ratio.

$9 \times 1.5 = 13.5$ ←

Multiply by the ratio to find the missing term.

The missing term is 13.5

Quick Test

1. Here are the first five terms of a sequence: $16, 12, 8, 4, 0 \ldots$
   Write down the next two terms.
2. The nth term of a sequence is $5n - 7$.
   Work out the 1st term and the 10th term of this sequence.
3. In the sequence below the next pattern is formed by adding another layer of tiles around the previous pattern.

   Work out how many tiles will be needed for the 6th pattern.

Key Words

sequence
term-to-term rule
position-to-term rule
nth term
recursive
arithmetic sequence
geometric sequence

Terms and Rules

You must be able to:

- Deduce and use expressions to calculate the nth term of a linear sequence
- Deduce and use expressions to calculate the nth term of a quadratic sequence
- Recognise and use sequences of triangular numbers, squares, cubes and other special sequences.

Finding the nth Term of a Linear Sequence

- A number sequence that increases or decreases by the same amount each time is called a **linear sequence**.
- To work out the expression for the nth term in a linear sequence, look for a pattern in the numbers.
- Using a function machine to represent a sequence of numbers can help.

The first five terms of a sequence are: 9, 12, 15, 18, 21 ... ◄

What is the expression for the nth term of this sequence?

Input (n)	× 3 ($3n$)	Output
1	3	9
2	6	12
3	9	15
4	12	18
5	15	21
n	$3n$	$3n + 6$

The difference between $3n$ and the output in each case is 6, so the expression for the nth term is **$3n + 6$**.

The term-to-term rule is +3, so the expression for the nth term starts with $3n$.

The 'input' is the position of the term and the 'output' is the value of the term.

> ### Key Point
>
> The zero term is the term that would come before the first term in a given sequence of numbers.

- The alternative method is to work out the **zero term**.

The first five terms of a sequence are: 20, 16, 12, 8, 4 ... ◄

What is the expression for the nth term of this sequence?

Input	Output
0	zero term
1	20
2	16
3	12
4	8
5	4
n	nth term

The expression for the nth ◄ term is $-4n + 24$ or $24 - 4n$.

The difference between terms is –4.

The zero term is (20 + 4 =) 24

The nth term = 'the difference' × n + the zero term

Special Sequences

- It is important to be able to recognise special sequences of numbers:
  - square numbers: 1, 4, 9, 16, 25 ... The nth term is n^2
  - cube numbers: 1, 8, 27, 64, 125 ... The nth term is n^3
  - triangular numbers: 1, 3, 6, 10, 15 ... The nth term is $\frac{n}{2}(n + 1)$
  - the Fibonacci sequence: 1, 1, 2, 3, 5, 8, 13, 21 ... The recursive relationship can be written as $U_{n+2} = U_{n+1} + U_n$

> Work out the expression for the nth term of the following sequence: 2, 9, 28, 65, 126 ...
>
> The expression for the nth term is $n^3 + 1$

The cube numbers produce the sequence 1, 8, 27, 64, 125 ... This sequence of numbers is one greater than the cube numbers.

Finding the nth Term of a Quadratic Sequence

- A **quadratic sequence** contains an n^2 term as part of the expression for the nth term.
- The general form of a quadratic sequence is $an^2 + bn + c$, where a, b and c are constants.
- To find the expression for the nth term of a quadratic sequence, look to see if you can spot the pattern.
- If this doesn't work, extend the difference backwards to get the zero term ($n = 0$).

> Work out the expression for the nth term of the following sequence of numbers: 0, 3, 8, 15, 24, 35 ...
>
> $n^2 - 1$

The numbers are one less than the square numbers.

> Work out the expression for the nth term of the following sequence of numbers: 4, 13, 26, 43 ...
>
> | n | 0 | 1 | 2 | 3 | 4 |
> |---|---|---|---|---|---|
> | c (nth Term) | −1 | 4 | 13 | 26 | 43 |
> | $a + b$ (First Difference) | | 5 | 9 | 13 | 17 |
> | $2a$ (Second Difference) | | | 4 | 4 | 4 |
>
> $c = -1$
> $2a = 4$, so $a = 2$ $a + b = 5$, so $b = 3$
> The expression for the nth term is $2n^2 + 3n - 1$

Set up a difference table that extends backwards to get the term for $n = 0$.

In the form $an^2 + bn + c$

Quick Test

1. a) Work out the expression for the nth term for the following sequence: 7, 10, 13, 16, 19 ...
   b) Work out the 50th term in this sequence.
2. Write down the next two terms in this quadratic sequence: −4, −1, 4, 11, 20, ___, ___
3. Write down an expression for the nth term for the following sequence of numbers: 0, 7, 26, 63, 124 ...

Key Words

linear sequence
zero term
Fibonacci sequence
quadratic sequence

Transformations

Geometry and Measures

You must be able to:

- Identify, describe and construct transformations of shapes, including reflections, rotations, translations and enlargements
- Describe the changes achieved by combinations of transformations.

Transformations

- **Reflection:**
  - Each point on the image is the same distance from the mirror line as the corresponding point on the object
  - The object and the image are **congruent** (same size and shape)
  - To define a reflection on a coordinate grid, the equation of the mirror line should be stated.
- **Rotation** is described by stating the:
  - Direction rotated (clockwise or anticlockwise)
  - Angle of rotation (which is usually a multiple of 90° in the exam)
  - Centre of rotation (point about which the shape is rotated).

> **Key Point**
>
> There is no need to state clockwise or anticlockwise for a rotation of 180°.

a) Describe the transformation that maps triangle ABC onto triangle $A'B'C'$.

   The transformation that maps ABC to $A'B'C'$ is a rotation of 90° anticlockwise (or 270° clockwise) about the origin (0, 0).

b) Describe the transformation that maps triangle $A'B'C'$ onto triangle $A''B''C''$.

   The transformation that maps $A'B'C'$ to $A''B''C''$ is a rotation of 180° about the origin (0, 0).

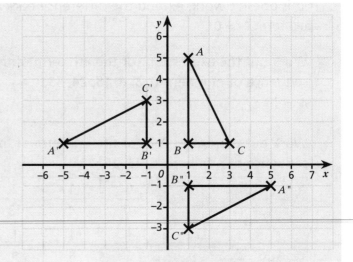

- **Translation:**
  - The shape does not rotate – it moves left or right and up or down – and stays the same size
  - The translation is represented by a **column vector** $\binom{x}{y}$.

Describe the transformation that takes shape A to shape B.

Four right, one up.

A translation through $\binom{4}{1}$

> **Key Point**
>
> x represents the distance moved **horizontally**: **positive** means to the **right** and **negative** means to the **left**.
>
> y represents the distance moved **vertically**: **positive** means **up** and **negative** means **down**.

- **Enlargement:**
  - The shape of the object is not changed, only its size. The enlarged shape is **similar** to the original shape
  - The **scale factor** determines whether the object gets bigger (scale factor > 1) or smaller (scale factor < 1)
  - Scale factors can be negative. This results in:
    - A 180° rotation of the object
    - The image being on the opposite side of the centre of enlargement to the object
  - When describing the enlargement, state the scale factor and the centre of enlargement.

Enlarge triangle A by scale factor 2, centre of enlargement (1, 2). Label the transformed triangle B.

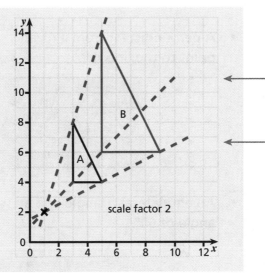

All construction lines must remain.

The side **lengths** of triangle B are **twice** the length of the corresponding sides of triangle A. However, the **area** of triangle B is **four times** bigger.

Combinations of Transformations

- It is possible for two or more transformations to be applied to a shape, one after the other.

Enlarge triangle C by scale factor $-\frac{1}{2}$ about centre of enlargement (0, 0) to form triangle D.

Transform triangle D by a reflection in the line $y = 1$ to form triangle E.

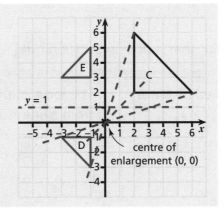

Quick Test

1. Describe the single transformation that takes:
   a) triangle A to triangle B
   b) triangle A to triangle C
   c) triangle A to triangle D.

Constructions

You must be able to:

- Use a ruler and a pair of compasses to produce different constructions, including bisectors
- Describe a locus and solve problems involving loci
- Understand and construct plans and elevations of 3D shapes.

Constructions

- The **perpendicular bisector** of line AB.

Open compasses to more than half AB. Put compass point on A. Draw arc. Put compass point on B. Draw arc. XY is the perpendicular bisector.

- A **perpendicular** from a given point to the line AB.

Put compass point on C. Draw arc. Keep radius the same. Put compass point on E. Draw arc. Put compass point on D. Draw arc. Join C to the point where the arcs cross.

- An **angle bisector**.

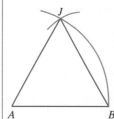

Put compass point on A. Draw arc FG. Put compass point on F. Draw arc. Put compass point on G. Draw arc. Join HA.

- An **equilateral triangle** and an **angle of 60°**.

Open compasses to length AB. Put point on A. Draw arc. Put compass point on B. Draw arc. Join AJ and BJ. Angle $A = 60°$

> **Key Point**
>
> The perpendicular distance from a point to a line is the shortest distance to the line.

> **Key Point**
>
> To construct a perpendicular at a given point on a line:
>
> Put the compass point on that point.
>
> Draw two arcs on the line either side of that point.
>
> Then construct the perpendicular bisector of the two new points.

Defining a Locus

- A **locus** is the path taken by a point that is obeying certain rules.
- The plural of locus is **loci**.

- The locus of points that are a **fixed distance from a given point** A is a circle.

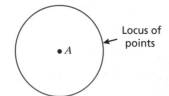

Locus of points

- The locus of points that are a **fixed distance from a line** AB.

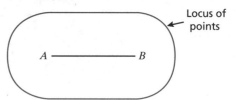

Locus of points

- The locus of points that are the **same distance from two lines** AB and AC. This is the angle bisector.

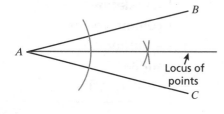

Locus of points

- The locus of points that are **equidistant**, or the **same distance, from two points** A and B. This is the perpendicular bisector.

Locus of points

Loci Problems

A guard dog is tied to a post by a 4-metre long rope.

Accurately draw the locus of the points the dog can reach using a scale of 1cm : 1m

The solution would be a shaded circle of radius 4cm.
The dog could reach the circumference of the circle and all points within it.

4cm   Not Drawn Accurately

Plans and Elevations

- The **plan view** of a 3D shape shows what it looks like from above – a bird's eye view.
- The side **elevation** is the view of a 3D shape from the side.
- The front elevation is the view from the front.

Here is a 3D shape made from centimetre cubes drawn on isometric paper.

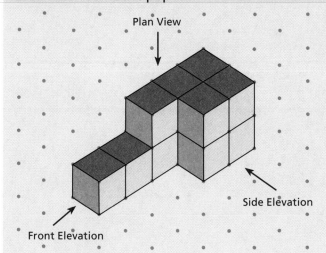

On squared paper, draw:

a) the plan view

b) the front elevation

c) the side elevation.

Quick Test

1. Describe how you would accurately construct an angle of 30°.
2. A rectangular vegetable plot *ABCD* measures 6m by 3m.
   A goat is tethered at corner *A* by a 4m rope.
   Accurately draw the plot and construct the locus of the points that the goat can reach. Scale 1cm : 1m. Shade the region of the vegetable plot that can be eaten by the goat.

Key Words

perpendicular
bisector
locus / loci
equidistant
plan view
elevation

Linear Graphs

You must be able to:

- Work with coordinates in all four quadrants
- Plot graphs of linear functions
- Work out the equation of a line through two given points or through one point with a given gradient
- Work out the gradient and y-intercept of a straight line in the form $y = mx + c$.

Drawing Linear Graphs from Points

- **Linear graphs** are straight-line graphs.
- The equation of a straight-line graph is usually given in the form $y = mx + c$, where m is the **gradient** of the line and c is the **intercept** of the y-axis.
- $y = mx + c$ is a function of x, where the input is the x-coordinate and the output is the y-coordinate.

> **Key Point**
>
> To draw a straight line, only two coordinates are needed.

Draw the graph of the equation $y = 2x + 5$

Use values of x from –3 to 3.

x	–3	0	3
y	–1	5	11

First, draw a table of values. Include a third value as a check.

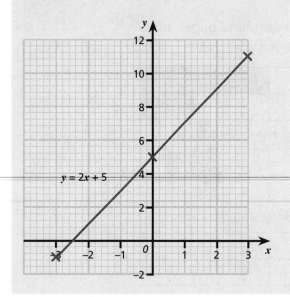

$y = 2x + 5$

Drawing Graphs by Cover-Up and Gradient–Intercept Methods

- Another method that can be used to draw a graph is the cover-up method.
- This can be used for equations in the form $ax + by = c$.

Draw the graph of the equation $2x + 3y = 6$

$3y = 6$
$y = 2$ ⟵ The y-intercept is (0, 2).

Cover up the x term and solve to find y.

$2x = 6$
$x = 3$ ⟵ The x-intercept is (3, 0).

Cover up the y term and solve to find x.

- The gradient–intercept method is used for equations in the form $y = mx + c$.

Draw the line with equation $y = 4x + 2$

The y-intercept is (0, 2). ⟵ Find the value for y when $x = 0$.

The gradient is 4. ⟵ m is the gradient in the equation $y = mx + c$

Therefore, another coordinate on the line is (1, 6). ⟵ Gradient $= \frac{4}{1}$ so for every 1 across, go up 4.

Finding the Equation of a Line

- To find the equation of a straight line in the form $y = mx + c$, work out the gradient and y-intercept.

Work out the equation of the line that joins the points (1, 20) and (4, 5).

Gradient $= \frac{15}{-3}$

$= -5$

$y = mx + c$

$y = -5x + c$ ⟵

$20 = (-5 \times 1) + c$

$c = 25$

The equation of the line is $y = -5x + 25$

To work out the value of c, substitute in point (1, 20) or (4, 5).

Quick Test

1. Draw the graph with equation $y = 4x + 1$
2. Write down the gradient and y-intercept of the line with equation $y = 5 - 2x$
3. Work out the equation of the line that joins the points (5, 7) and (3, 10).

Key Words

linear graph
gradient
intercept

Graphs of Quadratic Functions

You must be able to:

- Recognise, sketch and interpret graphs of quadratic functions
- Identify and interpret roots, intercepts and turning points of quadratic functions
- Work out roots using algebraic methods
- Work out turning points.

Plotting Quadratic Graphs

- A **quadratic equation** is an equation that contains an unknown term with a power of 2, e.g. x^2.
- You can use a table of values to draw **quadratic graphs**.

Draw the graph of the function $y = 2x^2 + 1$

x	−2	−1	0	1	2	3	4
y	9	3	1	3	9	19	33

$y = 2x^2 + 1$

The Significant Points of a Quadratic Curve

- A sketch shows the shape and significant points on a graph, but is not an accurate drawing.
- To sketch a quadratic, work out the **roots**, the **intercept** and the **turning point**, i.e. the **maximum** or **minimum point**.
- The roots are found by solving the quadratic when $y = 0$.
- Because quadratic curves are symmetrical, the x coefficient of the turning point is halfway between the two roots.

> **Key Point**
>
> All quadratic graphs have a line of symmetry, which passes through the turning point.
>
> The roots of a quadratic equation are the points where the graph crosses the x-axis. Not all quadratic curves will have roots.

Sketch the graph of equation $y = x^2 + 5x + 4$

Roots:

$x^2 + 5x + 4 = 0$

$(x + 4)(x + 1) = 0$

$x = -4$ or $x = -1$

y-intercept:

$y = 4$

y-intercept is (0, 4)

Substitute $x = 0$ into the equation.

Work out the values for x when $y = 0$.

Turning Point:

$x = -4$ and $x = -1$, therefore $x = -2.5$

$y = (-2.5)^2 + (5 \times -2.5) + 4$

$\quad = -2.25$

minimum point is (−2.5, −2.25)

The value of x for the turning point is in the middle of the two roots.

Substitute the value of x into the equation.

(−2.5, −2.25)

Transformation of the Graph $y = f(x)$

- Functions can be used to describe **transformations**.

Function	Transformation	
$f(x) + a$	Translation $\begin{pmatrix} 0 \\ a \end{pmatrix}$	Move up by a
$f(x) - a$	Translation $\begin{pmatrix} 0 \\ -a \end{pmatrix}$	Move down by a
$f(x + a)$	Translation $\begin{pmatrix} -a \\ 0 \end{pmatrix}$	Move left by a
$f(x - a)$	Translation $\begin{pmatrix} a \\ 0 \end{pmatrix}$	Move right by a
$-f(x)$	Reflection in x-axis	
$f(-x)$	Reflection in y-axis	

> **Key Point**
>
> If the coefficient of the x^2 term is **positive**, the graph will have a **minimum** point. If the coefficient of the x^2 term is **negative**, the graph will have a **maximum** point. The maximum and minimum point can also be found by completing the square (see p.97).

The curve $y = f(x)$ has a minimum point (2, −3).

Write down the coordinates of the turning point on the curve with equation:

a) $y = f(x + 2)$ The graph moves left by 2, so the minimum point is (0, −3).

b) $y = f(x) - 3$ The graph moves down by 3, so the minimum point is (2, −6).

c) $y = -f(x)$ The graph is reflected in the x-axis, so the maximum point is (2, 3).

> **Quick Test**
>
> 1. Sketch the graph of the equation $y = x^2 - 3x + 2$
> 2. Draw the graph with the equation $y = x^2 - 6$
> 3. A quadratic function $y = f(x)$ has a maximum at the point (5, 3). Write down the maximum of the curve with equation
>   **a)** $y = f(x) - 1$ and **b)** $y = f(x - 3)$

> **Key Words**
>
> quadratic equation
> quadratic graph
> roots
> intercept
> turning point
> maximum point
> minimum point
> transformation

Powers, Roots and Indices

You must be able to:

- Recognise and recall powers of 2, 3, 4 and 5
- Recognise and recall the square numbers up to 15×15
- Calculate with powers and roots, including fractional and negative indices
- Estimate powers and roots of positive numbers
- Carry out exact calculations with surds.

Roots and Powers (Indices)

- Powers or **indices** are a shorthand method of showing that a number is multiplied by itself a number of times, e.g.
  $5 \times 5 = 5^2$
  $5 \times 5 \times 5 = 5^3$
  $5 \times 5 \times 5 \times 5 = 5^4$, etc.
- A **root** is the inverse function of a power.
- You must learn all the square numbers up to 15×15, i.e.
  1, 4, 9, 16, 25, 36, 49, 64, 81, 100, 121, 144, 169, 196, 225.
- You must also learn the cubes of 1, 2, 3, 4, 5 and 10, i.e.
  1, 8, 27, 64, 125 and 1000.
- You must be able to recognise powers of 2, 3, 4 and 5, e.g.
  $16 = 2^4$ and $243 = 3^5$, and work out real roots.

Work out the value of 3^4.	Write down the value of $\sqrt[3]{8}$.
$3 \times 3 \times 3 \times 3 = 81$	$2^3 = 8$, so $\sqrt[3]{8} = 2$

Key Point

$x^0 = 1$

Estimating Powers and Roots

- Roots of positive integers are either integers, fractions or irrational.
- Roots that are not integers or fractions can be estimated.

Between which two positive integers does $\sqrt{40}$ lie?

$6^2 = 36$ and $7^2 = 49$, therefore $\sqrt{40}$ is between 6 and 7. ◄———

Key Point

A square root can be both positive or negative.

As a decimal $\sqrt{40} = 6.325$ (to 3 d.p.)

Multiplying and Dividing Powers

- When completing calculations involving powers, apply the following rules:

$$x^m \times x^n = x^{m+n}$$
$$x^m \div x^n = x^{m-n}$$
$$(x^m)^n = x^{mn}$$
$$\frac{1}{x^n} = x^{-n}$$

Write each of the following as a single power of 2.

a) $2^3 \times 2^4 = 2^{3+4}$
$= 2^7$

b) $2^9 \div 2^4 = 2^{9-4}$
$= 2^5$

c) $(2^3)^4 = 2^{3 \times 4}$
$= 2^{12}$

d) $\dfrac{1}{2^5}$
$= 2^{-5}$

Negative and Fractional Powers

- When working with negative and fractional powers, the same basic rules apply.
- Fractional powers are used to express roots:
  - $x^{\frac{1}{2}} = \sqrt{x}$
  - $x^{\frac{1}{3}} = \sqrt[3]{x}$
  - $x^{\frac{1}{n}} = \sqrt[n]{x}$
  - $x^{\frac{m}{n}} = \left(\sqrt[n]{x}\right)^m$

Key Point

$(\sqrt{x})^2 = x$

a) Write down the value of $49^{\frac{1}{2}}$

$49^{\frac{1}{2}} = \sqrt{49}$
$= 7$

b) Write down the value of $16^{\frac{3}{4}}$

$16^{\frac{3}{4}} = \left(\sqrt[4]{16}\right)^3$
$= 2^3$
$= 8$

Surds

- **Surds** are square roots that are not integers or fractions and are, therefore, irrational.
- Surds can be manipulated using the rules below:
  - $\sqrt{ab} = \sqrt{a} \times \sqrt{b}$
  - $\sqrt{\dfrac{a}{b}} = \dfrac{\sqrt{a}}{\sqrt{b}}$

Key Point

To keep calculations exact, it is sometimes best to leave your answer in surd form.

Simplify $\sqrt{200}$.

$\sqrt{200} = \sqrt{100 \times 2}$
$= \sqrt{100} \times \sqrt{2}$ Always make sure one of the numbers is a square number.
$= 10\sqrt{2}$

- It is bad practice to leave a surd on the denominator of a fraction.
- To find an equivalent fraction, **rationalise** the denominator.

Rationalise $\dfrac{1}{\sqrt{3}}$

$\dfrac{1}{\sqrt{3}} \times \dfrac{\sqrt{3}}{\sqrt{3}} = \dfrac{\sqrt{3}}{3}$ Multiply the numerator and denominator by $\sqrt{3}$.

Key Words

index / indices
root
surd
rationalise

Quick Test

1. Simplify $2x^3 \times 3x^2$
2. Write down the value of $25^{\frac{3}{2}}$
3. Rationalise $\dfrac{2}{\sqrt{2}}$

Area and Volume 1

You must be able to:

- Recall and use the formulae for the circumference and area of a circle
- Recall and use the formula for the area of a trapezium
- Recall and use the formulae for the volume and surface area of a prism
- Recall and use the formulae for the volume and surface area of a cylinder.

Circles

LEARN

Circumference of a Circle (C) = $2\pi r$ or $C = \pi d$

Area of a Circle (A) = πr^2

Key Point

The symbol π represents the number **pi**.

π can be approximated to 3.14 or $\frac{22}{7}$

Work out the circumference and area of a circle with radius 9cm. Give your answers to 1 decimal place.

Circumference

$C = 2 \times \pi \times 9$

$\quad = 18 \times \pi$

$\quad = 56.5$cm (to 1 d.p.)

Area

$A = \pi \times 9^2$

$\quad = \pi \times 81$

$\quad = 254.5$cm^2 (to 1 d.p.)

Trapeziums

LEARN

The area of a **trapezium** is:

$A = \frac{1}{2}(a + b)h$

where a and b are the **parallel** sides and h is the **perpendicular** height

Key Point

Perpendicular means 'at right angles'.

Parallel means 'in the same direction and always the same distance apart'.

- This formula can be proved:

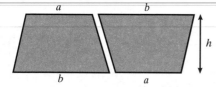

- Two identical trapeziums fit together to make a parallelogram with base $a + b$ and height h.
- The area of the parallelogram is $(a + b)h$.
- Therefore, the area of each trapezium is $\frac{1}{2}(a + b)h$.

Work out the area of the trapezium.

$A = \frac{1}{2} \times (5 + 10) \times 4$

$\quad = 30$cm^2

Key Point

The area of a parallelogram is: $A = bh$

Prisms

- A right prism is a 3D shape that has the same **cross-section** running all the way through it.

> **Volume of a Prism = Area of Cross-Section × Length**

- The surface area is the sum of the areas of all the **faces**.

Work out the volume and surface area of the triangular prism.

Volume   Surface Area

Area of the cross-section
$$= \tfrac{1}{2} \times 3 \times 4 = 6\text{cm}^2$$
Volume $= 6 \times 7$
$$= 42\text{cm}^3$$

Five faces:
Two triangular faces = 6 + 6 = 12
Base = 4 × 7 = 28
Side = 3 × 7 = 21
Slanted side = 5 × 7 = 35
Total surface area:
12 + 28 + 21 + 35 = 96cm²

Cylinders

> **Volume of a Cylinder = $\pi r^2 h$**
>
> **Surface Area of a Cylinder = $2\pi rh + 2\pi r^2$**

Key Point

A cylinder is just like any other right prism. To find the volume, you multiply the area of the cross-section (circular face) by the length of the cylinder.

Work out the volume and surface area of the cylinder. Give your answers in terms of π.

Volume
$V = \pi \times 4^2 \times 7$
$= 112\pi\text{cm}^3$

Surface Area
$SA = 2 \times \pi \times 4 \times 7 + 2 \times \pi \times 4^2$
$= 56\pi + 32\pi$
$= 88\pi\text{cm}^2$

> **Quick Test**
>
> 1. Calculate the volume and surface area of a cylinder with radius 4cm and height 6cm.
> 2. Work out the area of the trapezium.
>
>
>
> 3. Calculate the circumference and area of a circle, diameter 7cm.

Key Words

trapezium
parallel
perpendicular
cross-section
face

Area and Volume 2

You must be able to:

- Find the volume of a pyramid
- Find the volume and surface area of a cone
- Find the volume of a frustum
- Find the volume and surface area of a sphere
- Find the area and volume of composite shapes.

Pyramids

- A **pyramid** is a 3D shape in which lines drawn from the **vertices** of the base meet at a point.

LEARN   Volume of a Pyramid = $\frac{1}{3}$ × Area of the Base × Height

> **Key Point**
>
> A pyramid is usually defined by the base, e.g. a square-based pyramid or a triangular-based pyramid.

Work out the volume of the square-based pyramid.

$V = \frac{1}{3} \times 9 \times 9 \times 7$

$= 189\,\text{cm}^3$

7cm

9cm   9cm

Cones

- A **cone** is a 3D shape with a circular base that tapers to a single vertex.

LEARN   Volume of a Cone = $\frac{1}{3}\pi r^2 h$

Surface Area of a Cone = $\pi r l + \pi r^2$

Work out **a)** the volume and **b)** the surface area of the cone.

Give your answers to 1 decimal place.

6cm

4cm

a) $h = \sqrt{6^2 - 4^2}$ ← First find the height using Pythagoras' Theorem.

$h = \sqrt{20}$

$V = \frac{1}{3} \times \pi \times 4^2 \times \sqrt{20} = 74.9\,\text{cm}^3$

b) $SA = (\pi \times 4 \times 6) + (\pi \times 4^2) = 125.7\,\text{cm}^2$

- A **frustum** is the 3D shape that remains when a cone is cut parallel to its base and the top cone removed.
- The original cone and the smaller cone that is removed are always similar.

LEARN

$$\text{Volume of a Frustum} = \text{Volume of Whole Cone} - \text{Volume of Top Cone}$$

Calculate the volume of the frustum. Leave your answer in terms of π.

Radius of small cone = 3cm

$$V = \frac{1}{3}\left(\pi \times 6^2 \times 8\right) - \frac{1}{3}\left(\pi \times 3^2 \times 4\right)$$

$$= 84\pi\,\text{cm}^3$$

> The two cones are similar with scale factor 2.

Spheres

- A **sphere** is a 3D shape that is round, like a ball. At every point, its surface is equidistant from its centre.

> Volume of a Sphere = $\frac{4}{3}\pi r^3$
>
> Surface Area of a Sphere = $4\pi r^2$

- A **hemisphere** is half of a sphere; a dome with a circular base.

Work out **a)** the volume and **b)** the surface area of the sphere. Leave your answers in terms of π.

a) $V = \frac{4}{3} \times \pi \times 6^3 = 288\pi\,\text{cm}^3$

b) $SA = 4 \times \pi \times 6^2 = 144\pi\,\text{cm}^2$

Composite Shapes

Calculate the area of the shaded region.

$$A = (6 \times 7) - (\pi \times 1.5^2)$$

$$= 34.9\,\text{cm}^2 \text{ (to 1 d.p.)}$$

> Find the area of the rectangle and subtract the area of the circle.

Key Point

To find the volume of a composite shape, you must break the shape down.

Work out the volume of the shape.

Give your answer in terms of π.

Volume of the cylinder = $2.5^2 \times \pi \times 7.8 = \dfrac{195}{4}\pi\,\text{cm}^3$

Volume of the cone = $\frac{1}{3} \times \pi \times 2.5^2 \times 6.2 = \dfrac{155}{12}\pi\,\text{cm}^3$

Total volume = $\dfrac{195}{4}\pi + \dfrac{155}{12}\pi = \dfrac{185}{3}\pi\,\text{cm}^3$

Quick Test

1. Work out the volume of a sphere with diameter 10cm.
2. Calculate the surface area of a cone with radius 3cm and perpendicular height 6cm.
3. Work out the volume of a square-based pyramid with side length 5cm and perpendicular height 8cm.
4. Calculate the surface area of a hemisphere with radius 6cm.

Key Words

pyramid
vertex / vertices
cone
frustum
sphere
hemisphere

Review Questions

Order and Value

1 An adult theatre ticket costs £38.60 and a child ticket costs £12.76.

Work out the total cost for 12 adult tickets and 5 child tickets. [2]

2 If $1.263 \times 2.47 = 3.11961$, work out:

  a) 12.63×0.247 [1]

  b) 0.1263×0.247 [1]

3 What is the value of $(-3)^5$? [1]

> **Total Marks** _____ / 5

Types of Number

1 Use algebra to find three consecutive numbers that add up to 111. [3]

2 2, 9, 13, 15, 27, 35, 100

From the list of numbers above, find:

  a) A factor of 72. [1]

  b) A multiple of 7. [1]

  c) A square number. [1]

  d) A cube number. [1]

  e) A prime number. [1]

3 Find a number (other than 1) that is a square number and also a cube number. [1]

4 $25^2 \times 25^3 = 25^m$

Work out the value of m. [2]

5 Peter says $(2^3)^2 = (2^2)^3$

Is Peter right or wrong? Explain your answer. [2]

> **Total Marks** _____ / 13

Basic Algebra & Factorisation and Formulae

1 There are k children in a room.

The number of children who wear glasses is g.

Write an expression in terms of k and g for the number of children who **do not** wear glasses. [1]

2 Simplify $7x - 2y + 5x - 3y$ 🖩 [2]

3 Work out the value of the following expression when $a = -3$. 🖩

$$\frac{4a^2 - a^3}{a^4}$$ [1]

4 Expand and simplify $(y + 3)(2y - 7)(6 - y)$ 🖩 [3]

5 Factorise $5ab - 3b^2c$ 🖩 [1]

6 Explain why $5p - 7q$ cannot be factorised. 🖩 [1]

7 Solve $\frac{x}{3} + 3 = 1$ 🖩 [1]

8 The shape below is a rectangle.

6x + 5

3x + 2

Mitan thinks the correct expression for the perimeter of the rectangle is $9x + 7$.

a) Mitan is wrong. Explain his mistake. [1]

b) The perimeter of the rectangle is 56cm. Work out the value of x. [2]

9 The formula for the volume of a cylinder is $V = \pi r^2 h$.

a) Work out the volume of a cylinder with a radius (r) of 2cm and a height (h) of 10cm. [1]

b) Make r the subject of the formula. [2]

c) Work out the value of r when $V = 50$ and $h = 10$. [2]

Total Marks _____ / 18

Review Questions

Ratio and Proportion

1. The square of the speed (*v*) at which a ball is thrown is directly proportional to the height (*h*) reached. A ball thrown at a speed of 10 metres per second reaches a height of 5 metres.

   Calculate the height reached by a ball thrown at a speed of 30 metres per second. [4]

2. £60 is divided in the ratio 5 : 7

   What is the difference in value between the two shares? [3]

3. Simplify 6.2 hours : 4 minutes [2]

   Total Marks _____ / 9

Variation and Compound Measures

1. Martin is investing £400. He can choose between two ways of saving:

   Type A – Simple interest at 6% per annum

   Type B – Compound interest at 5% per annum

   Which savings plan will give the better return after four years?
   You **must** show your working. [4]

2. a) Calculate the distance travelled by a mouse moving at 1.5 metres per second for 1.5 seconds. [2]

   b) Misty, a farm cat, can run at 11.25 miles per hour. Misty chases a mouse moving at 1.5 metres per second into a straight, plastic pipe.

   If the mouse enters the pipe two seconds before the cat and the pipe is six metres long, would Misty catch the mouse before it escapes out of the other end? You **must** show your working. Assume 5 miles = 8 kilometres. [3]

3. It takes three dogs 15 days to eat a large sack of dog biscuits.

   How long would it take five dogs to eat the same sized sack of dog biscuits at the same rate? [2]

   Total Marks _____ / 11

Angles and Shapes 1 & 2

1 The three interior angles of a triangle are $y°$, $2y°$ and $3y°$.

Work out the size of the largest angle.   [2]

2 A quadrilateral has one angle of 80°. Another angle is twice as big as the first angle and a third angle is 20° smaller than the first angle.

Work out the size of the fourth angle.   [2]

3 This angle diagram is incorrect. Explain why.   [1]

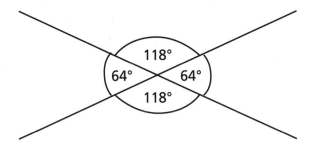

4 An aircraft flies from airport A on a bearing of 054° to airport B.

Work out the bearing that the aircraft must follow in order to return to airport A.   [1]

5 Work out the exterior angle of a regular 15-sided shape. [2]

6 An irregular hexagon has interior angles in degrees of $2h$, $4h$, $4h$, $4h$, $5h$ and $5h$.

Work out the size of the smallest angle. [3]

7 State whether each of the following statements is **true** or **false**.

a) The sum of the interior angles of a heptagon is 900°. [1]

b) A parallelogram has no lines of symmetry and rotational symmetry of order 2. [1]

c) The direction south-east is on a bearing of 145°. [1]

Total Marks _____ / 14

Review Questions

Fractions

1 Work out $\frac{4}{25} + \frac{3}{5} \times \frac{2}{5}$ 📱 [2]

2 How many nails of length $\frac{8}{9}$ cm can be placed end to end on a 32cm line? 📱 [1]

3 What is half of a half of a quarter? 📱 [1]

4 A box contained 344 dog biscuits.
Fluff ate one-eighth of them.

How many biscuits were left? 📱 [2]

Total Marks _____ / 6

Percentages

1 A coat costs £56. It is reduced by 35% in a sale.

What is the sale price of the coat? [2]

2 Richard says that 30% of £40 is the same as 40% of £30.

Is he correct? Explain your answer. [2]

3 A house is purchased for £215 000 and sold for £300 000.

Calculate the percentage profit. [1]

4 On Black Friday, a coat was reduced by 35%.

If the sale price of the coat on Black Friday is £80, what was its original price to the nearest pound? [2]

Total Marks _____ / 7

Probability 1 & 2

1   The sides of a spinner are coloured red, blue, green and yellow.
The probability that the spinner will land on each colour is shown in the table below:

Colour	Red	Blue	Green	Yellow
Probability	0.3	0.3		0.1

    a) Complete the table to show the probability of landing on green. [1]

    b) Estimate the number of times the spinner will land on the colour red if it is spun 50 times. [2]

    c) Georgia and Annie are playing a game using the spinner.
Georgia suggests that she wins if the spinner lands on red or blue and Annie wins if the spinner lands on green or yellow.

      Georgia thinks this is fair. Is she correct? Explain your answer. [2]

2   A restaurant serves three courses: starters, mains and desserts. All customers have a main course. The Venn diagram shows information about customers who also had a starter, a dessert or both.

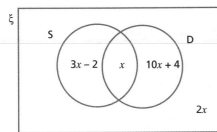

S = Starter
D = Dessert

    The total number of customers served at lunchtime is 82.

    Work out the number of customers who ordered neither a starter nor a dessert. [3]

3   Two events, A and B, are mutually exclusive: $P(A) = 0.3$ and $P(B) = 0.6$

    a) Draw a Venn diagram to represent this information. [2]

    b) Write down $P(A \text{ and } B)$ [1]

    c) Work out $P(A \text{ or } B)$ [2]

4   Helen has 10 yoghurts. Four are vanilla flavoured and six are banana flavoured.
She takes a yoghurt at random for breakfast on Monday and another on Tuesday.

    Work out the probability that she takes one of each flavour. [4]

Total Marks .... / 17

Number Patterns and Sequences & Terms and Rules

1   Here is a sequence:

20, 16, 12, 8, 4 ...

Write down the expression for the nth term of the sequence. [2]

2   Work out the first five terms in the sequence with the nth term $3n - 5$. [2]

3   The first five terms of an arithmetic sequence are 14, 17, 20, 23, 26 ...

a)   Write an expression for the nth term of this sequence. [2]

b)   Calculate the 100th term in this sequence. [1]

4   The population of a culture of bacteria is given by the formula:
$P = 10 \times 2^t$, where t is the time in hours.

a)   What is the initial population of bacteria? [2]

b)   What is the size of the population after five hours? [2]

c)   After how many hours will the population exceed 1000? [2]

d)   Is this a sensible formula to model the population? Explain your answer. [1]

5   **a)**   A sequence of numbers is given as 5, 8, 12, 19 ...

Write down the next two terms in the sequence. [2]

b)   A second sequence of numbers is given as 2, 3, 2, 3, 2, 3 ...

Write down the 100th term. [1]

6   These are the first two patterns in a sequence made using matchsticks:

a)   Draw the next two patterns in the sequence. [2]

b)   Write down the rule for the number of matchsticks required for pattern number n. [2]

c)   Use the rule to work out how many matchsticks are required for pattern 100. [1]

7   The first five terms of a quadratic sequence are:

4, 15, 32, 55, 84 ...

Write down the expression for the nth term of this sequence. [3]

Total Marks .... / 25

Transformations & Constructions

1 a) Plot the following points: $A(2, 0)$ $B(5, 0)$ $C(5, 2)$ $D(3, 2)$ $E(3, 5)$ $F(2, 5)$
     Join the points together and label the shape M.   [1]

   b) Rotate shape M by 180° about the origin to form shape N.   [1]

   c) Reflect shape N in the x-axis to form shape O.   [1]

   d) Describe fully the single transformation that maps shape O to shape M.   [2]

2 Rectangle R has a width of 3cm and a length of 5cm. It is enlarged by scale factor 3 to give rectangle T.

   a) What is the area of rectangle T?   [2]

   b) How many times bigger is the area of rectangle T than the area of rectangle R?   [2]

3 Describe how to construct an angle of 45°.   [2]

4 Describe the locus of points in the following:

   a) A person sitting on the London Eye as it rotates around.   [1]

   b) The seat of a moving swing.   [1]

   c) The end of the minute hand on a clock moving for one hour.   [1]

   d) The end of a moving see-saw.   [1]

5 The diagram represents a solid made from 10 identical cubes.
   On a squared grid, draw the:

   a) Front elevation   [1]

   b) Plan view.   [1]

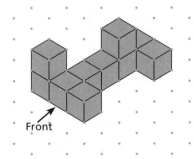
Front

Total Marks _____ / 17

Linear Graphs & Graphs of Quadratic Functions

1 Draw the graph of $y = 4x - 2$ with values of x from –4 to 4. [2]

2 Draw the graph of the function with gradient 5 and y-intercept (0, 3). [2]

3 Write down the gradient and y-intercept of the graph with equation $y = 5 - 2x$. [1]

4 Draw the graph with equation $y = 3x^2 - 2x + 1$ for values of x from –3 to 3. [2]

5 Work out the equation of the line that joins the points (–2, 5) and (3, –1). [3]

6 Work out the equation of the line drawn. [3]

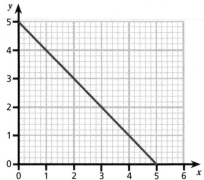

7 A straight line goes through the points (T, S) and (H, R).
$S = R - 5$ and $T = H + 2$

   a) What is the gradient of the line? [2]

   b) What is the y-intercept in terms of H and R? [1]

8 The graph has a minimum value at
the point (0, –8) and a maximum value at
the point (–2, –4).

Work out the new minimum and maximum
values after the following transformations.

   a) $f(x + 2)$ [2]

   b) $f(-x)$ [2]

   c) $f(x) + 4$ [2]

   d) $f(x - 5)$ [2]

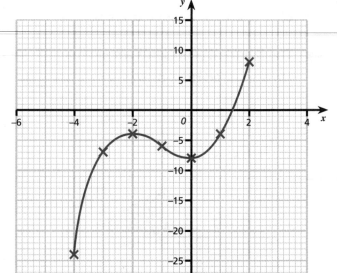

Total Marks _____ / 24

Powers, Roots and Indices

1 Write down the following as a single power of 5:

    a) $5^2 \times 5^3$ [1]

    b) $5^7 \div 5^3$ [1]

    c) $\dfrac{5^4 \times 5^2}{5^3}$ [2]

2 Write down the value of $16^{-\frac{1}{2}}$.   [2]

3 Simplify $(x^2)^4$ [1]

4 Write the following in the form $a\sqrt{3} + b$ where a and b are integers.

$$\left(2 + \sqrt{3}\right)\left(2 + 2\sqrt{3}\right)$$
    [3]

5 Simplify $\left(3r^2 p\right)^3$ [2]

> **Total Marks** .... / 12

Area and Volume 1 & 2

1 Two identical circles sit inside a square of side length 6cm.

  Work out the area of the shaded region.   [4]

2 A vase is made from two cylinders. The larger cylinder has a radius of 15cm.
The total volume of the vase is $6000\pi\,\text{cm}^3$.
The ratio of volumes of the smaller cylinder to the larger cylinder is 1 : 3

    a) Calculate the height of the larger cylinder.   [3]

    b) The height and radius of the smaller cylinder are equal. Work out the radius of the
        smaller cylinder.   [3]

3 A cat's toy is made out of plastic. The top of the toy is a solid cone with radius 3cm and
height 7cm. The bottom of the toy is a solid hemisphere. The base of the hemisphere and the
base of the cone are the same size.

  Calculate the volume of plastic needed to make the toy. Give your answer in terms of π.   [3]

> **Total Marks** .... / 13

Uses of Graphs

You must be able to:

- Use the form $y = mx + c$ to identify parallel and perpendicular lines
- Interpret the gradient of a straight-line graph as a rate of change
- Recognise and interpret graphs that illustrate direct and inverse proportion.

Parallel and Perpendicular Lines

- **Parallel** lines travel in the same direction and have the same gradient.
- **Perpendicular** lines are at right angles to each other.
- If a line has gradient m, then any line perpendicular to it will have a gradient of $-\frac{1}{m}$.

Write down the gradient of the line parallel to the line with the equation $y = 7 - 2x$.

The line has a gradient of –2 so the line that is parallel to it also has a gradient of –2.

> **Key Point**
>
> The gradient of a straight line in the form $y = mx + c$ is m.

Write down the gradient of the line that is perpendicular to the line with equation $y = 3x - 1$.

The line has a gradient of 3 so the line that is perpendicular to it has a gradient of $-\frac{1}{3}$.

Work out the equation of the line that goes through the point (2, 9) and is parallel to the line with equation $y = 7x + 10$.

$y = mx + c$

$y = 7x + c$ ← Substitute in $m = 7$

$9 = (7 \times 2) + c$ ← Goes through the point (2, 9), so $x = 2$ when $y = 9$

$c = -5$

The equation of the parallel line is $y = 7x - 5$.

Work out the equation of the line that is perpendicular to the line $y = \frac{3}{2}x + 2$ and goes through the point (10, 6).

$y = mx + c$

$y = -\frac{2}{3}x + c$ ← The line has a gradient of $\frac{3}{2}$ so the gradient of the perpendicular line is $-\frac{2}{3}$. This is your value for m.

$6 = (-\frac{2}{3} \times 10) + c$ ← Goes through the point (10, 6), so $x = 10$ when $y = 6$

$c = \frac{38}{3}$

The equation of the perpendicular line is
$y = -\frac{2}{3}x + \frac{38}{3}$ or $3y = 38 - 2x$

Gradient of a Line

- The **rate of change** is the rate at which one quantity changes in relation to another.
- The gradient of a straight-line graph represents a rate of change – it describes how the variable on the y-axis changes when the variable on the x-axis is increased by 1.

The graph below shows the volume of liquid in a container over time. What is the rate of change?

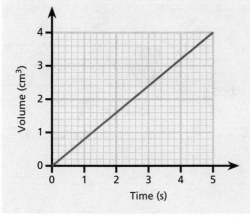

Gradient = $\frac{4}{5}$

$= \frac{4}{5} = 0.8\text{cm}^3/\text{s}$

Gradient = $\frac{\text{Change in } y}{\text{Change in } x}$

The gradient is the rate of change.

Real-Life Uses of Graphs

The graph below is the conversion graph between miles and kilometres.

a) How many kilometres are there in 5 miles?
   5 miles = 8km

Read from the graph.

b) What is the gradient of the line?

Gradient = $\frac{8}{5}$

$= 1.6$

1 mile = 1.6km

Graph: Kilometres (y-axis, 0–10) vs Miles (x-axis, 0–5)

Other Graphs 1

You must be able to:

- Recognise, draw and interpret cubic, reciprocal and exponential graphs
- Interpret distance–time graphs and velocity–time graphs
- Work out acceleration from a velocity–time graph
- Work out speed from a distance–time graph.

Distance–Time Graphs

- A **distance–time graph** shows distance travelled in relation to a fixed point (starting point) over a period of time.
- The gradient of a straight line joining two points is the speed of travel between those two points.

Key Point

Speed = $\dfrac{\text{Distance}}{\text{Time}}$

The graph below shows Val's car journey from St Bees to Cockermouth and back.

a) Val sets off at 12 noon and travels directly to Cockermouth. At what time does she arrive?

Val travels for 30 minutes so arrives at 12.30pm.

b) For how long does Val stop in Cockermouth?

20 minutes ◄──── This is represented by the horizontal line on the graph – where the distance does not change.

c) Val begins her journey home but stops to fill up with petrol. Calculate the average speed of Val's journey from the petrol station to home in kilometres per hour.

Speed = $\dfrac{\text{Distance}}{\text{Time}} = \dfrac{25}{0.5}$ ◄──── Convert minutes into hours.

= 50km/h

Velocity–Time Graphs

- **Velocity** has both magnitude (size) and direction.
- The magnitude of velocity is called speed.
- The gradient of a straight line joining two points is the acceleration between those two points.
- Area Under Graph = Distance Travelled

The graph below shows part of the journey of a car.

a) For how many seconds does the car decelerate?

4 seconds

b) What is the distance travelled in the last 5 seconds of the journey?

5 × 0.8 = 4m

c) What is the acceleration of the car in the first 3 seconds of the journey?

The acceleration is $\frac{1}{3}$ m/s²

Revise

Key Point

A positive gradient shows increasing speed.

A negative gradient shows decreasing speed.

Zero gradient shows constant speed.

Between 6 seconds and 10 seconds there is a negative gradient, so the car is decelerating.

This is the area of the rectangle under that part of the line.

This is the gradient of the line.

Other Graphs

- A **cubic function** is one that contains an x^3 term.
- A **reciprocal** function takes the form $y = \frac{a}{x}$
- An **exponential** function takes the form $y = k^x$

Cubic Function

Reciprocal Function

Exponential Function

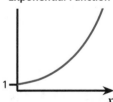

Quick Test

1. Plot the graph $y = 4^x$ for x values –4 to 4.
2. Plot the graph $y = 3x^3 - 5$ for values –2 to 2.
3. Below is a graph for the journey of a car.

a) What is the total distance travelled?
b) For how many minutes is the car decelerating?

Key Words

distance–time graph
velocity
cubic function
reciprocal
exponential

Other Graphs 2

You must be able to:

- Estimate and interpret the area under a curve
- Work out and interpret a gradient at a point on a curve
- Work out the equation of a tangent to a circle.

Estimating the Area Under a Curve

- On a velocity–time graph, the area under the curve is equal to the distance travelled.
- To estimate the area under a curve, split the area into triangles and trapeziums. Remember, this is only an estimate.

Below is the velocity–time graph for a journey.

Estimate the distance travelled.

The area of the triangle on the left is $\frac{1}{2} \times 2 \times 20 = 20$

The area of the trapezium is $\frac{1}{2} \times (20 + 15) \times 1 = 17.5$

The area of the triangle on the right is $\frac{1}{2} \times 1 \times 15 = 7.5$

Look carefully to check the units on the axes match, i.e. time in hours and speed in distance per hour.

The total area is 20 + 17.5 + 7.5 = 45

An estimate of the total distance travelled is 45 miles.

Rates of Change

- The **gradient** of a straight line describes how the variable on the y-axis changes when the variable on the x-axis is increased by 1 unit.
- On a distance–time graph, speed is the change in distance per unit of time. This is called a rate of change.
- The gradient of a curve is constantly changing.
- The gradient at any particular point is found by drawing a **tangent** at that point.

Work out the gradient of the curve $y = x^2$ at the point where $x = 3$.

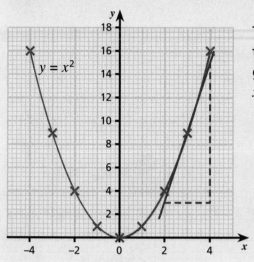

The gradient of the tangent is $\frac{12}{2} = 6$, so the gradient of the curve at $x = 3$ is 6.

Draw a tangent that meets the curve at the specified point.

Equation of a Circle

- The equation of a circle is of the form $x^2 + y^2 = r^2$ where the centre is (0, 0) and the radius is r.
- A circle with equation $x^2 + y^2 = 25$ has centre (0, 0) and radius 5.

Work out the equation of the tangent to the circle with equation $x^2 + y^2 = 25$ at the point (3, 4).

The gradient of the radius is $\frac{4}{3}$.

The gradient (m) of the perpendicular line, i.e. the tangent, is $-\frac{3}{4}$.

$4 = (-\frac{3}{4} \times 3) + c$

$c = \frac{25}{4}$

Substitute $x = 3$, $y = 4$ and $m = -\frac{3}{4}$ into $y = mx + c$ to work out c.

The equation of the tangent is $y = -\frac{3}{4}x + \frac{25}{4}$ OR $4y + 3x = 25$

Quick Test

1. Work out the equation of the tangent of the circle with equation $x^2 + y^2 = 29$ at the point (2, 5).
2. Plot the curve with equation $y = 3x^2 + 5$ and estimate the gradient at the point (2, 17).

Key Words

gradient
tangent

Inequalities

Algebra

You must be able to:

- Solve linear inequalities in one or two variables
- Solve quadratic inequalities in one variable
- Represent solutions to inequalities on number lines or graphs.

Linear Inequalities

- The solution to an **inequality** can be shown on a number line.

 means $x <$ means $x >$

●——← means $x \leqslant$ ●——→ means $x \geqslant$

Solve these inequalities and show the solutions on a number line:

a) $x + 3 > 4$

$x > 4 - 3$

$x > 1$

```
  -1   0   1   2   3
```

b) $2(x + 4) \leqslant 18$

$x + 4 \leqslant 9$

$x \leqslant 5$

```
  0   1   2   3   4   5   6
```

Work out all the possible integer values of n for these inequalities:

a) $-4 < n < 4$

$n = -3, -2, -1, 0, 1, 2, 3$

b) $-3 < 10n \leqslant 53$

$-0.3 < n \leqslant 5.3$ ←

$n = 0, 1, 2, 3, 4, 5$ ←

Divide each part of the inequality by 10.

n must be a whole number.

Graphical Inequalities

- The graph of the equation $y = 6$ is a line; the graph of the inequality $y > 6$ is a **region**, which has the line $y = 6$ as a boundary.
- For inequalities $>$ and $<$ the boundary line is **not included** in the solution and is shown as a **dashed line**.
- For inequalities $\geqslant$ and $\leqslant$ the boundary line is **included** in the solution and is shown as a **solid line**.

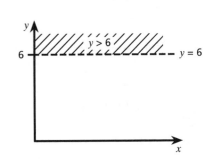

On a graph, show the region that satisfies $x \geqslant 0$, $x + y < 3$ and $y > 3x - 1$.

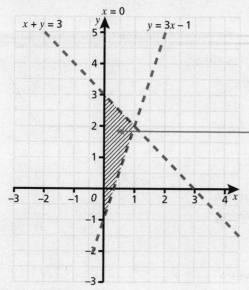

This is the region that satisfies all three inequalities.

Quadratic Inequalities

- A quadratic inequality contains an x^2 term and has a maximum of two possible solutions for x.
- To solve a quadratic inequality you must factorise the side with the quadratic.
- To determine the direction of the arrow for each solution, substitute values for x in the pairs of brackets.

Solve $x^2 + 2x - 15 \geqslant 0$ and show the solution on a number line.

$(x + 5)(x - 3) \geqslant 0$ ←   Factorise the expression.

If $(x + 5)(x - 3) = 0$, then $x = -5$ or 3

$(x + 5)(x - 3) \geqslant 0$ ←   Substitute values for x to determine the direction of the arrow.

If $x = -10$ then $(-10 + 5)(-10 - 3) = 65$, which satisfies $\geqslant 0$, so the arrow goes to the left from -5.

If $x = 10$, then $(10 + 5)(10 - 3) = 105$, which satisfies $\geqslant 0$, so the arrow goes to the right from 3.

So the solutions are $x \geqslant 3$ and $x \leqslant -5$.

Quick Test

1. Solve the inequality $2x - 5 < 9$ and show the solution on a number line.
2. Solve the inequality $5x + 4 < 3x + 10$
3. Solve $x^2 + 5x - 14 \geqslant 0$

Key Words

inequality
region

Congruence and Geometrical Problems

You must be able to:

- Identify congruent and similar shapes
- State the criteria that congruent triangles satisfy
- Solve problems involving similar figures
- Understand geometrical problems.

Congruent Triangles

- If two shapes are the same size and shape, they are **congruent**.
- Two triangles are congruent if they satisfy one of the following four criteria:
  - SSS – three sides are the same
  - SAS – two sides and the **included angle** (the angle between the two sides) are the same
  - ASA – two angles and one corresponding side are the same
  - RHS – there is a right angle, and the hypotenuse and one other corresponding side are the same.
- Sometimes angles or lengths of sides have to be calculated before congruency can be proved.

State whether these two triangles are congruent and give a reason for your answer.

Angle CAB = Angle PQR (given)

$AC = QR$ (given)

$AB = PQ$ (given)

Triangles ABC and PQR are congruent because they satisfy the criteria SAS.

> **Key Point**
>
> Congruent shapes can be reflected, rotated or translated and remain congruent.

Similar Triangles

- **Similar** figures are identical in shape but can differ in size.
- In similar triangles:
  - corresponding angles are identical
  - lengths of corresponding sides are in the same ratio $y : z$
  - the area ratio $= y^2 : z^2$
  - the volume ratio $= y^3 : z^3$.

Triangles AED and ABC are similar.

Calculate **a)** AC and **b)** DC.

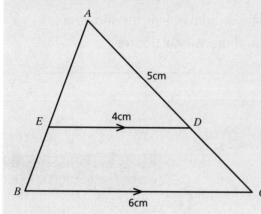

a) $\dfrac{4}{6} = \dfrac{5}{AC}$ ← The corresponding sides of both triangles are in the same ratio.

$4 \times AC = 6 \times 5$ ← Cross multiply.

$AC = \dfrac{6 \times 5}{4}$

$= 7.5\text{cm}$

b) $DC = AC - AD$

$= 7.5 - 5$

$= 2.5\text{cm}$

Geometrical Problems

- Congruency and similarity are used in many geometric proofs.

Prove that the base angles of an isosceles triangle are equal.

Given $\triangle ABC$ with $AB = AC$
Let M be the midpoint of BC
Join AM
$AB = AC$ (given)
$BM = MC$ (from construction)
$AM = AM$ (common side)
$\triangle ABM$ and $\triangle ACM$ are congruent (SSS)
So, angle $ABC =$ angle ACB

Quick Test

1.

a) Prove triangle BAC is similar to triangle DAE.
b) Calculate the length of BC.

Key Words

congruent
included angle
similar

Right-Angled Triangles

You must be able to:

- Recall and use the formula for Pythagoras' Theorem
- Recall the trigonometric ratios
- Use the trigonometric ratios to calculate unknown lengths and angles in right-angled triangles in two and three-dimensional figures.

Pythagoras' Theorem

- **Pythagoras' Theorem states** $a^2 + b^2 = c^2$.
- The longest side (c) is opposite the right angle and called the **hypotenuse**.

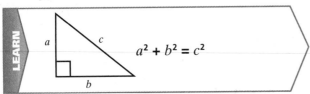

$$a^2 + b^2 = c^2$$

> Careful: b is a shorter side so the answer must be **less** than 14.1cm.

> Unless stated otherwise, give the answer to 3 significant figures.

Work out the length of the hypotenuse (c).

$3^2 + 4^2 = c^2$
$9 + 16 = c^2$
$c = \sqrt{25}$
$c = 5m$

Work out the length of b.

$14.1^2 - 7.2^2 = b^2$
$198.81 - 51.84 = b^2$
$146.97 = b^2$
$b = \sqrt{146.97}$
$b = 12.1cm$

A boat sails 15km due north, then 10km due east.

How far is the boat from its starting point? Give your answer to 3 decimal places.

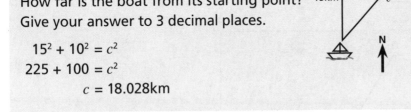

$15^2 + 10^2 = c^2$
$225 + 100 = c^2$
$c = 18.028km$

- The following length combinations are known as Pythagorean triples and regularly appear in right-angled triangles:
  - (3, 4, 5)
  - (6, 8, 10)
  - (5, 12, 13)
  - (7, 24, 25).
- Memorise them to help identify unknown lengths quickly.

> **Key Point**
>
> You need to know the exact values of the trigonometric ratios for common angles:
>
> $\sin 0° = 0$
>
> $\cos 0° = 1$
>
> $\tan 0° = 0$
>
> $\sin 30° = \frac{1}{2}$
>
> $\cos 30° = \frac{\sqrt{3}}{2}$
>
> $\tan 30° = \frac{1}{\sqrt{3}}$
>
> $\sin 45° = \frac{1}{\sqrt{2}}$
>
> $\cos 45° = \frac{1}{\sqrt{2}}$
>
> $\tan 45° = 1$
>
> $\sin 60° = \frac{\sqrt{3}}{2}$
>
> $\cos 60° = \frac{1}{2}$
>
> $\tan 60° = \sqrt{3}$
>
> $\sin 90° = 1$
>
> $\cos 90° = 0$
>
> You might find it useful to learn the decimal values too.

ABCDEFGH is a cuboid.
Calculate the length of *AH* to 2 decimal places.

This is a Pythagorean triple, so
EH = 5m.

$$6^2 + 5^2 = c^2$$
$$36 + 25 = c^2$$
$$61 = c^2$$
$$c = \sqrt{61} = 7.81\text{m}$$
So *AH* = 7.81m

Key Point

a^2 means $a \times a$ **not** $2 \times a$.

Trigonometric Ratios

- You can calculate unknown sides or angles in right-angled triangles using **sine**, **cosine** and **tangent**.
- θ is a Greek letter called **theta**. It stands for the unknown angle.

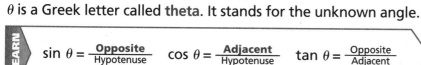

$$\sin \theta = \frac{\text{Opposite}}{\text{Hypotenuse}} \quad \cos \theta = \frac{\text{Adjacent}}{\text{Hypotenuse}} \quad \tan \theta = \frac{\text{Opposite}}{\text{Adjacent}}$$

- The above formulae can be remembered by:

Some **O**ld **H**orses **C**arry **A** **H**eavy **T**on **O**f **A**pples
SOH   **CAH**   **TOA**

- For example, $\sin \theta$ (**S**ome) $= \dfrac{\text{Opposite (Old)}}{\text{Hypotenuse (Horses)}}$

Key Point

Sine, cosine and tangent ratios can **only** be used in right-angled triangles.

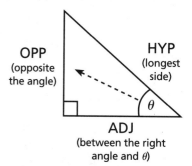

OPP
(opposite the angle)

HYP
(longest side)

ADJ
(between the right angle and θ)

Work out *x* to 1 decimal place.

$$\sin 40° = \frac{x}{7}$$
$$x = 7 \times \sin 40°$$
$$x = 7 \times 0.6428 = 4.499...$$
$$x = 4.5\text{cm}$$

In this example you are looking for a missing length.

Work out θ to the nearest degree.

$$\tan \theta = \frac{8}{5} = 1.6$$
$$\tan^{-1} 1.6 = 57.99°$$
$$\theta = 58°$$

In this example you are looking for a missing angle.

On the calculator press either
SHIFT tan 1 . 6 or
2ndF tan 1 . 6

Key Words

Pythagoras' Theorem
hypotenuse
sine
cosine
tangent
theta
opposite
adjacent

Quick Test

1. An isosceles triangle has side lengths of 7cm, 7cm and 5cm. Calculate the angle between the two equal sides to the nearest degree.
2. A triangle has sides 8cm, 15cm and 17cm. Is it right-angled?
3. *ABC* is a right-angled triangle, where angle *B* = 90°. *AB* = 6cm, *AC* = 9cm. Calculate *BC* to 2 decimal places.

Sine and Cosine Rules

You must be able to:

- Recall and use the sine rule to work out an unknown side or angle of a triangle
- Recall and use the cosine rule to work out an unknown side or angle of a triangle
- Recall and use Area $= \frac{1}{2}ab \sin C$ to calculate the area, sides or angles of a triangle.

Solving Any Triangle

- The **sine rule** is used to calculate unknown angles or side lengths in triangles that are not right-angled.

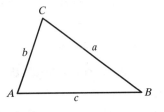

> **Key Point**
>
> Capital letters are used to represent angles; lower case letters are used to represent sides.

LEARN

Sine Rule: $\dfrac{a}{\sin A} = \dfrac{b}{\sin B} = \dfrac{c}{\sin C}$

Calculate the size of angle C.

$$\frac{6}{\sin 40°} = \frac{7}{\sin C}$$

$6 \times \sin C = 7 \times \sin 40°$ ← Rearrange.

$\sin C = \dfrac{(7 \times \sin 40°)}{6}$

$\sin C = \dfrac{4.499513268}{6}$

$\sin C = 0.749918878$

$C = \sin^{-1} 0.749918878$ ← Always check your calculator is in degree mode.

$C = 48.58°$

Calculate the length of side BC to 2 decimal places.

$$\frac{a}{\sin 71°} = \frac{8}{\sin 50°}$$ ← BC is side a.

$a \times \sin 50° = 8 \times \sin 71°$

$a = \dfrac{(8 \times \sin 71°)}{\sin 50°}$

$a = \dfrac{(8 \times 0.9455185756)}{0.7660444431}$

$a = \dfrac{7.564148605}{0.7660444431}$

$a = 9.87$cm (to 2 d.p.)

- The **cosine rule** is used for triangles without a right angle when:
  - two sides and the included angle (between them) are known
  - three sides are known.

LEARN

Cosine Rule: $a^2 = b^2 + c^2 - 2bc \cos A$ **OR** $\cos A = \dfrac{\left(b^2 + c^2 - a^2\right)}{2bc}$

Calculate the size of angle A to 1 decimal place.

$$\cos A = \frac{(b^2 + c^2 - a^2)}{2bc}$$

$$\cos A = \frac{(5^2 + 10^2 - 9^2)}{2 \times 5 \times 10}$$

$$\cos A = \frac{44}{100}$$

$$\cos A = 0.44$$

Angle $A = \cos^{-1} 0.44$

Angle $A = 63.9°$

Calculate the length of side BC to 2 decimal places.

$a^2 = b^2 + c^2 - 2bc \cos A$ ⟵ $\boxed{BC \text{ is side } a.}$

$a^2 = 9^2 + 12^2 - 2 \times 9 \times 12 \times \cos 47°$

$a^2 = 225 - 216 \times \cos 47°$

$a^2 = 225 - 147.31$

$a^2 = 77.69$

$a = \sqrt{77.69}$

$a = 8.81\text{cm}$ (to 2 d.p.)

Using Sine to Calculate the Area of a Triangle

- This formula for calculating the area of any triangle can be rearranged to work out unknown sides and angles:

$$\text{Area} = \tfrac{1}{2}ab \sin C = \tfrac{1}{2}bc \sin A = \tfrac{1}{2}ac \sin B$$

Key Point

If Area $= \frac{1}{2}ab \sin C$, then

$$a = \frac{2 \times \text{Area}}{b \sin C} \text{ and}$$

$$b = \frac{2 \times \text{Area}}{a \sin C}$$

$$\sin C = \frac{2 \times \text{Area}}{ab}$$

The area of the triangle ABC is 35.27cm². Calculate the length of AC. Give your answer to the nearest whole number.

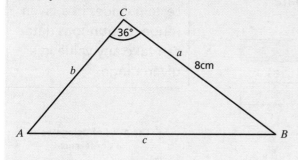

$\text{Area} = \frac{1}{2}ab \sin C$ ⟵ $\boxed{\text{Rearrange to make } b\ (AC) \text{ the subject.}}$

$$\frac{2 \times \text{Area}}{a \sin C} = b$$

$$\frac{2 \times 35.27}{8 \times \sin 36°} = b$$

$$\frac{70.54}{4.7023} = b$$

$$b = AC = 15\text{cm}$$

Quick Test

1. In triangle ABC, angle $A = 44°$, angle $C = 64°$ and $BC = 4$cm. Calculate **a)** the length of AB and **b)** the area of triangle ABC to 3 decimal places.

Key Words

sine rule
cosine rule

Statistics 1

You must be able to:

- Understand how a sample can be used to represent a population and its limitations
- Understand and identify different types of data
- Interpret and construct appropriate tables, charts and diagrams for data
- Calculate the mean, median, mode and range of a set of data.

Sampling

- A **population** is a collection of individuals or items.
- For some populations, collecting information from all members is not practical, so information is obtained from a **sample**.
- There are advantages and limitations associated with sampling:

Advantages	Limitations
Quicker and cheaper than investigating whole population	Bias can occur
Can be impractical to investigate whole population	Different samples could give different results

- There are two main types of sample: **random** and **stratified**.
- In a stratified sample, the population is divided into subgroups. The sample includes members from all subgroups in the same proportions as they occur in the population.
- **Primary data** is collected by yourself or on your behalf.
- **Secondary data** is collected from a different source.

Statistical Representation

- Data can be categorised as **qualitative** (non-numerical) or **quantitative** (numerical).
- Numerical data can be **discrete** or **continuous**.
- The type of diagram you use will be influenced by the data itself and what aspect of the data you want to look at.

Chart	Purpose
Pie chart	To show proportion but not exact frequencies
Bar chart	To compare frequencies
Line graph	To show changes in trends over time

Statistical Measures

- To compare data sets you should compare:
  - a measure of average: **mean**, **median** or **mode**
  - a measure of spread, i.e. the **range** – the difference between the largest and smallest value in the data set.
- A stem and leaf diagram orders data from smallest to largest and can be used to find the median and range of discrete data.

Key Point

Limitations can be minimised by ensuring the sample is large enough and representative of the whole population.

Key Point

In a stratified sample, the sample size for each subgroup is calculated using the formula:

$$\frac{\text{Size of Subgroup}}{\text{Size of Population}} \times \text{Sample Size}$$

Key Point

Discrete data takes certain values in a given range. Continuous data can take any value in a given range.

The Ages of Delegates at a Conference

2	1 1 2 3 4 4
3	2 2 2 3 5 8
4	3 3 5 8
5	1 5 6

Key: 2 | 1 = 21 years

The table shows the times taken by students, in minutes, to complete a mathematical puzzle.

Time Taken to Complete Puzzle (mins)	Frequency (f)	Cumulative Frequency	Midpoint (x)	fx
$0 < t \leqslant 2$	8	8	1	8
$2 < t \leqslant 4$	12	20	3	36
$4 < t \leqslant 6$	10	30	5	50
$6 < t \leqslant 8$	5	35	7	35
	$\sum f = 35$			$\sum fx = 129$

a) Write down the modal class.
The modal class is $2 < t \leqslant 4$.

b) Work out which group contains the median.
$\dfrac{35 + 1}{2} = 18$th value
The median (18th value) is in the group $2 < t \leqslant 4$.

c) Estimate the mean time taken.
Mean $= \dfrac{129}{35} = 3.69$ minutes (to 2 d.p.)

Scatter Diagrams

- **Scatter diagrams** are used to investigate the relationship between two variables.
- If a linear relationship exists, a line of best fit can be drawn. This can be used to make predictions.
- A prediction taken from the line of best fit within the data range is reliable. Taken from outside the data range, it may not be reliable.
- **Positive correlation** – as one variable increases, the other variable increases.
- **Negative correlation** – as one variable increases, the other variable decreases.
- You can describe correlation as weak or strong.

Positive Correlation

As temperature increases, ice-cream sales increase

Negative Correlation

As distance travelled increases, petrol remaining decreases

Quick Test

1. The table below shows the age of children who attend a reading group at the library. Calculate **a)** the mean **b)** the median **c)** the mode and **d)** the range of the data.

Age	6	7	8	9
Frequency	5	10	6	12

2. Write down one advantage and one disadvantage of taking a sample instead of surveying the whole population.

Statistics 2

You must be able to:

- Construct and interpret diagrams for grouped data, including frequency polygons, histograms and cumulative frequency curves
- Construct and interpret box plots
- Compare data sets using quartiles and interquartile range.

Frequency Polygons

- A **frequency polygon** can be used to represent **grouped data**.
- A point is plotted for the midpoint of each group and joined by straight lines.

Cumulative Frequency Graphs

- The **cumulative frequency** is the running total. It is found by adding all the frequencies together.
- A cumulative frequency curve can be used to estimate the **median** and **interquartile range**.

Draw a cumulative frequency curve to represent the following data. Use your graph to estimate the median and interquartile range.

Time Taken to Run 100m (s)	Frequency	Cumulative Frequency
$12 < t \leqslant 14$	4	4
$14 < t \leqslant 17$	21	(4 + 21 =) 25
$17 < t \leqslant 19$	14	(25 + 14 =) 39
$19 < t \leqslant 23$	9	(39 + 9 =) 48

Work out the cumulative frequency and plot against the value at the end of the group.

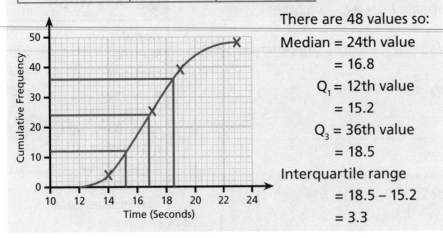

There are 48 values so:

Median = 24th value
        = 16.8

Q_1 = 12th value
      = 15.2

Q_3 = 36th value
      = 18.5

Interquartile range
        = 18.5 – 15.2
        = 3.3

Key Point

The lower **quartile** is the value $\frac{1}{4}$ or 25% of the way through the data set. The upper quartile is $\frac{3}{4}$ or 75% of the way through the data set.

The lower quartile is called Q_1

The upper quartile is called Q_3

The interquartile range is $Q_3 - Q_1$

Histograms and Box Plots

- A **histogram** is used to represent **continuous data**.
- The area of each bar on a histogram is proportional to the frequency for that class.
- The height of the bar is equal to the frequency density.

Construct a histogram to represent the data in the previous example for time taken to run 100m.

Time Taken to Run 100m (s)	Frequency	Class Width	Frequency Density
$12 < t \leqslant 14$	4	2	(4 ÷ 2 =) 2
$14 < t \leqslant 17$	21	3	(21 ÷ 3 =) 7
$17 < t \leqslant 19$	14	2	(14 ÷ 2 =) 7
$19 < t \leqslant 23$	9	4	(9 ÷ 4 =) 2.25

Key Point

Frequency Density
$= \dfrac{\text{Frequency}}{\text{Class Width}}$

- A **box plot** displays information about the range, median and quartiles of a data set. It must be drawn against a scale.

Draw a box plot for the data for the time taken to run 100m.

Use the values taken from the cumulative frequency graph.

Quick Test

1. For the following data construct a histogram.

Length of Leaf (cm)	Frequency
$5 < l \leqslant 7$	9
$7 < l \leqslant 12$	17
$12 < l \leqslant 20$	23

2. The box plot below shows information about the length of worms.
   a) Write down the median.
   b) Work out an approximate value for the interquartile range.

Key Words

frequency polygon
cumulative frequency
median
interquartile range
quartile
histogram
box plot

Review Questions

Number Patterns and Sequences & Terms and Rules

1 The first term that the following two sequences have in common is 17.

8, 11, 14, 17, 20 …

1, 5, 9, 13, 17 …

Work out the next term that the two sequences have in common.
You must show your working. [2]

2 Regular pentagons of side length 1cm are joined together to make a pattern.

a) Use the patterns to complete the table below.

Pattern Number	Perimeter (cm)
1	
2	
3	
4	
60	
n	

[2]

b) What is the maximum number of pentagons that could be used to give a perimeter
less than 1500cm? [2]

3 The nth term of a sequence is $n^2 + 6$.
Jenny states that 'the nth term is odd if n is odd and the nth term is even if n is even'.

Is Jenny correct? Explain your answer. [2]

4 Write down the nth term for the sequence below:

4, 6, 10, 18, 34 … [1]

Total Marks .... / 9

Transformations & Constructions

1 Three points $X(5,1)$, $Y(3, 5)$ and $Z(1, 2)$ are reflected in the y-axis.

   a) Give the new coordinates of the three points. [3]

   b) The original points X, Y and Z are rotated 90° about $(0, 0)$ in a clockwise direction.

     Give the coordinates of the three points in their new positions. [3]

2 A rectangle (C) measures 3cm by 5.5cm. Each length of rectangle C is enlarged by scale factor 3 to form a new rectangle (D).

   What is the ratio of the area of rectangle C to rectangle D? [3]

3 Describe the locus of points for the following:

   a) The path of a rocket for the first three seconds after take-off. [1]

   b) A point just below the handle on an opening door. [1]

   c) The central point of a bicycle wheel as the bicycle travels along a level road. [1]

   d) The end of a pendulum on a grandfather clock. [1]

4 Describe the plan view of a cube measuring 4cm by 4cm by 4cm. [1]

5 A can of baked beans has a circular lid of circumference 22cm and a height of 8cm.

   Draw the side elevation of the tin when it is standing upright. (Take π as $\frac{22}{7}$) [2]

6 The photograph shows a World War II Lancaster Bomber.

Sketch:

   a) The side elevation of the Lancaster Bomber [2]

   b) The front elevation of the Lancaster Bomber [2]

   c) The plan view of the Lancaster Bomber. [2]

Total Marks _____ / 22

Linear Graphs & Graphs of Quadratic Functions

1 Work out the equation of the line that joins the points $\left(\frac{2}{3}, 8\right)$ and $\left(\frac{5}{6}, 5\right)$. [3]

2 Work out the equation of the line drawn below. [3]

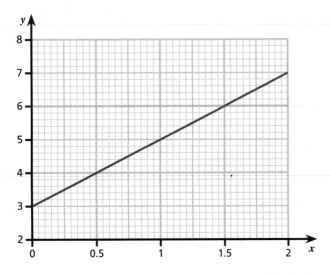

3 Sketch the graph of the function $y = x^2 + 4x + 3$, clearly stating the roots and the coordinates of the turning point. [3]

4 The equation of a line is $4y = 3x + 1$

   Work out the gradient and y-intercept of the line. [2]

5 A curve has the equation $y = x^2 + ax + b$
   The curve crosses the x-axis at the points (–7, 0) and (1, 0).

   a) Work out the values of a and b. [3]

   b) Work out the coordinates of the turning point. [2]

6 a) Sketch the graph $y = \frac{1}{x}$ [1]

   b) On the same axes sketch the graph $y = -\frac{1}{x}$ [1]

   c) Describe two different transformations that will transform the graph in part **a)** to the graph in part **b)**. [2]

Total Marks _____ / 20

Powers, Roots and Indices

1 Simplify $\left(2ab^{-5}\right)^{-3} \times \left(3a^{-2}b^3\right)^2$ [3]

2 Write $\frac{2}{\sqrt{3}} - \frac{6}{\sqrt{27}} + 6\sqrt{48}$ in the form $a\sqrt{3}$, where a is an integer. [3]

3 Amber states that $(x^{-2})^3 = \frac{1}{x^6}$

Is Amber's statement true or false? Explain your answer. [2]

4 Work out the value of $25^{-\frac{1}{2}}$. [1]

5 $a < \sqrt{170} < b$, where a and b are integers. $b - a = 1$.

Work out the values of a and b. [2]

Total Marks .... / 11

Area and Volume 1 & 2

1 **a)** Work out the volume of the triangular prism. [2]

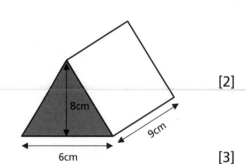

b) A cube has the same volume as the triangular prism.

Work out the total length of all the edges of the cube. [3]

2 The numerical values of the area and circumference of a circle are equal.

Work out the radius of this circle. [2]

3 The volume of the trapezoid is 900cm³.
All measurements are in centimetres.

Work out the value of x. [4]

4 The surface area of a sphere is 75cm². Work out the length of the radius. [3]

Total Marks .... / 14

Uses of Graphs & Other Graphs 1 & 2

1 A line is parallel to the line of equation $y = 3x - 2$ and goes through the point (1, 5).

Work out the equation of the line. [3]

2 Complete the table below.

Gradient of Line	Gradient of Parallel Line	Gradient of Perpendicular Line
5		
	3	
		$-\frac{1}{4}$
$\frac{8}{9}$		

[3]

3 Gemma, Naval and Esmai entered a five-mile cycling race.
The graph below shows the race.

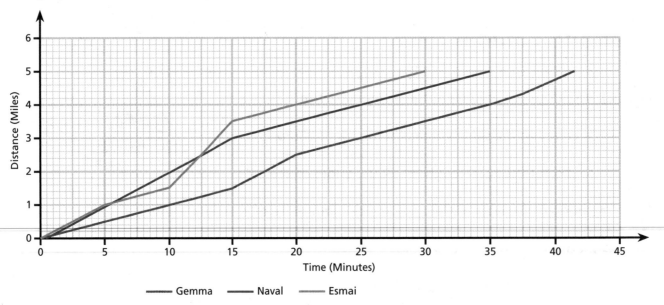

a) Who won the race? [1]

b) What speed was Naval travelling at for the last 20 minutes before he finished?
Give your answer in miles per hour. [2]

c) Between what times was Gemma travelling her fastest?
Give a reason for your answer. [2]

d) How many minutes after the race started did the winner move into the lead? [1]

e) Describe the race. [3]

Uses of Graphs & Other Graphs 1 & 2 (Cont.)

4 **a)** On the same set of axes, sketch the graphs of $y = 3^x$ and $y = 5^x$. [2]

b) Write down one property that the two graphs have in common. [1]

5 The graph below shows the conversion rate between British Pound Sterling (£) and the US Dollar ($).

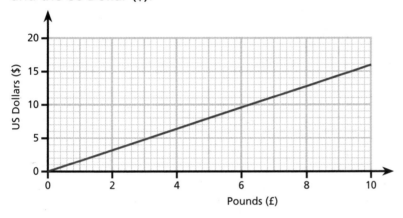

Pounds (£)

a) Calculate the gradient of the line. [2]

b) Interpret the gradient in terms of the exchange rate between British Pounds Sterling and US Dollars. [1]

6 The circle has equation $x^2 + y^2 = 41$

a) Work out the equation of the tangent to the circle at the point (4, –5). [4]

b) Write down the equation of a line parallel to the tangent. [1]

7 The graph below shows the journey of a car.

Work out the total distance travelled. [3]

Time (Seconds)

Total Marks .... / 29

Inequalities

1 Solve $5 - 4x > 25$ and show the answer on a number line. [2]

2 Write down the inequality represented by the shaded region.

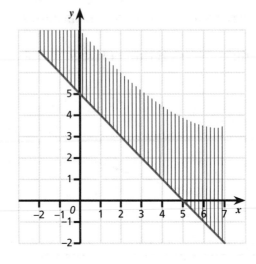

[1]

3 A bus can carry no more than 52 passengers.

Write down an inequality equation to show this information, where p is the number of passengers. [2]

> **Total Marks** .... / 5

Congruence and Geometrical Problems

1 Prove that triangle ABC and triangle BCD are similar. [3]

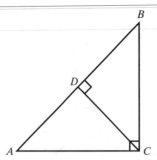

2 The corresponding lengths of two cuboids are 12cm and 3cm.

What is the ratio of their volumes? [1]

> **Total Marks** .... / 4

Right-Angled Triangles

1. A man walks 6.7km due north. He then turns due west and walks 7.6km.

   How far is he now from his starting point? [3]

2. A 4m ladder leans against a vertical wire fence. The foot of the ladder is 2m from the base of the fence. Fang, the lion, can jump 3m vertically.

   Will Fang be able to jump over the fence?
   You must give reasons for your answer. [4]

3.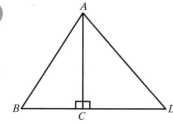
   AB = 10cm, BC = 8cm and CD = 6cm.

   a) Calculate AD to 3 significant figures. [3]

   b) What type of triangle is triangle ABD? [1]

4. A bumblebee leaves its nest and flies 10 metres due south and then 6 metres due west.

   What is the shortest distance the bumblebee has to fly to return to its nest?
   Give your answer to 3 significant figures. [3]

5. A triangle has side lengths of 1.5cm, 2.5cm and 2cm. Is it a right-angled triangle?
   Explain your answer. [3]

6. How long is the diagonal of a square of side length 3cm? [2]

7. A regular hexagon is inscribed in a circle of radius 6cm.

   Use sine to prove that the side length of the hexagon is 6cm. [2]

8. A is the point (4, 0) and B is the point (7, 5).

   Calculate the angle between AB and the x-axis to the nearest degree. [2]

9. A lifeguard is at the top of a lookout tower of height 14m, situated on a small island.
   She sees a swimmer (P) due west of her at an angle of depression of 35°.
   She sees another swimmer (Q) due south at an angle of depression of 18°.
   Work out:

   a) The distance of swimmer P from the base of the tower. [2]

   b) The distance of swimmer Q from the base of the tower. [2]

   c) The shortest distance between swimmers P and Q. Give your answer to 1 decimal place. [2]

Total Marks .... / 29

Sine and Cosine Rules

1. In a triangle ABC, AC = 5.7cm and AB = 7.5cm. The area of the triangle is 20cm^2.

   Calculate angle CAB to the nearest degree. [4]

2. ABC is a triangle where BC = 12cm and CA = 14cm.

   If angle ABC = 50°, calculate angle BAC to 2 decimal places. [3]

3. A tourist was standing 30m from the base of the Leaning Tower of Pisa.
   The angle of elevation to the top of the tower was 30°.
   The distance of the tourist from the top of the tower was 80m.

   Calculate the slanting height (l) of the Leaning Tower of Pisa.
   Give your answer to the nearest metre. [5]

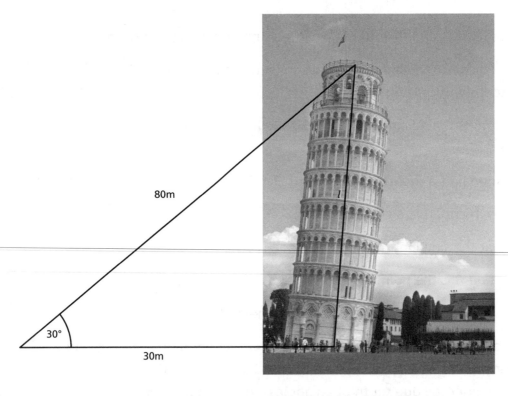

4. In a golf championship, the distance from the tee to the third hole is 250m.
   A golfer hits the ball 220m and it ends up 78m from the hole.

   Calculate the angle (from the tee) that the ball went off course to 2 decimal places. [2]

Total Marks _____ / 14

Statistics 1 & 2

1 Below is a table that shows the times taken for a group of students to run 100m.

Time (seconds)	Frequency
$10 < t \leqslant 11$	6
$11 < t \leqslant 12$	21
$12 < t \leqslant 14$	13
$14 < t \leqslant 18$	10

a) How many students ran 100m? [1]

b) Construct a histogram for this data. [3]

c) Use your histogram to estimate the number of students who took between 11 and 15 seconds to complete the run. [2]

2 The cumulative frequency curve gives information about the test results for 150 students. The lowest mark was 10 and the highest was 75.

Use this information to draw a box plot to represent the data. [4]

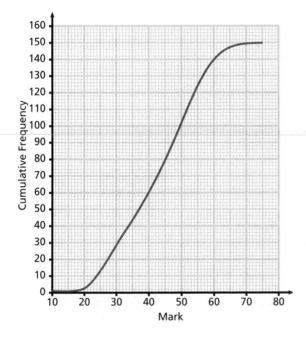

3 The table below shows hours of sunshine and amount of rainfall for nine towns across England in one month.

Sunshine (h)	600	420	520	630	470	380	560	430	450
Rainfall (mm)	11	18	13	9	16	25	14	20	19

a) Draw a scatter diagram to represent this information. [2]

b) Describe the relationship between hours of sunshine and rainfall. [1]

Total Marks .... / 13

Measures, Accuracy and Finance

You must be able to:

- Understand and solve problems relating to household finance
- Check calculations using approximation
- Round numbers and measures to an appropriate degree of accuracy
- Apply and interpret limits of accuracy, including for questions set in context.

Solving Real-Life Problems

- Being able to understand and solve problems relating to finance is an essential life skill.

> Nadine bought a designer handbag from eBay® for £230 and sold it two years later for £184.
>
> Calculate the percentage loss.
>
> Percentage profit or loss = $\dfrac{\text{profit or loss}}{\text{original amount}} \times 100$ Loss = £230 − £184 = £46
>
> $= \dfrac{£46}{£230} \times 100 = 20\%$

> VAT is charged at the standard rate of 20%.
>
> What is the final cost after VAT has been added to a bill of £45?
>
> $\dfrac{20}{100} \times £45 = £9$
>
> Total bill = £45 + £9
>
> $= £54$ This could also be calculated using a multiplier: 1.2 × £45 = £54

> Yvonne earns £36 000 per year. The first £8200 is tax free, but income tax of 20% must be paid on the rest.
>
> How much income tax does Yvonne have to pay each year?
>
> £36 000 − £8200 = £27 800   Work out the amount that is taxable.
>
> Income tax = $\dfrac{20}{100} \times £27\,800$
>
> $= £5560$

Limits of Accuracy

- You will often be asked to give answers to a certain number of **decimal places** or **significant figures**. To do this, you must 'round off' the number.
- **Decimal places** refers to the number of digits after the decimal point, e.g.

  - 27.3652 = 27.4 (to 1 d.p.)   Because the digit two places after the decimal point is 5 or more, the 3 (in the first decimal place) rounds up to 4.
  - 27.3652 = 27.37 (to 2 d.p.)   The 5 rounds up the 6 to a 7.

- The first **significant figure** is the first non-zero figure (working from left to right), e.g.
  - 541 = 540 (to 2 significant figures)
  - 0.0347 = 0.035 (to 2 significant figures).

Approximation of Calculations

- You can use **approximations** to estimate and check the answers to calculations.

> Estimate the answer to (2136 + 39.7) ÷ (9.6 × 11.1)
>
> (2000 + 40) ÷ (10 × 10) ⟵
>
> = 2040 ÷ 100
>
> = 20.4

Key Point

Leading zeros in decimals are **not** counted when rounding to significant figures.

Approximate each number to 1 significant figure.

Problems Involving Limits of Accuracy

- Measurement can be approximate, e.g. a length (x) of 145cm to the nearest centimetre could be anything from 144.5cm to 145.49999… cm
- Inequality notation is used to specify simple error intervals due to rounding, e.g. $144.5 \leqslant x < 145.5$
- When numbers are rounded off, the actual value lies between a **lower bound** and an **upper bound**. The bounds are called the **limits of accuracy**.

> There are 300 passengers (p), to the nearest hundred, on a train. Write down an inequality to show the possible values of p.
>
> $250 \leqslant p < 350$ ⟵

It would also be true to say $250 \leqslant p \leqslant 349$

> A tennis court is measured and is found to be 23.4m by 10.2m. Each measurement is given to the nearest 10cm.
>
> Calculate the maximum and minimum possible values for the area of the tennis court.
>
> Maximum area = 23.45 × 10.25 = 240.3625m² ⟵
>
> Minimum area = 23.35 × 10.15 = 237.0025m² ⟵

Multiply the upper bounds.

Multiply the lower bounds.

Key Words

decimal places
significant figures
approximation
lower bound
upper bound
limits of accuracy

Quick Test

1. Estimate the value of $(0.897)^2 \times 392.4$
2. A rectangle measures 2.81cm by 3.87cm to 2 decimal places. Write down the limits of accuracy in which each dimension lies.

Solving Quadratic Equations

You must be able to:

- Solve quadratic equations by factorising
- Solve quadratic equations by using a graph
- Solve quadratic equations by using the quadratic formula
- Solve quadratic equations by completing the square.

Factorisation

- When solving a quadratic equation by **factorisation** (see p.14–15), make sure it equals zero first.

> Solve the equation $x^2 + 4x + 3 = 0$ by factorisation.
>
> | × | x | +1 |
> |---|---|---|
> | x | x^2 | $+x$ |
> | +3 | $+3x$ | +3 |
>
> $(x + 1)(x + 3) = 0$
> $x + 1 = 0$ $x + 3 = 0$
> $x = -1$ $x = -3$

Key Point

If two brackets have a product of zero, one of the brackets must equal 0.

Set up and complete a table. The missing terms need to have a product of +3 and a sum of +4.

First row = first bracket; first column = second bracket

The Method of Intersection

- Plotting a graph of a quadratic equation can give zero, one or two solutions for x when $y = 0$.
- The solutions are given by the curve's points of **intersection** with the x-axis. These solutions will sometimes be estimates.

Key Point

The points of intersection with the x-axis are called roots.

> Find approximate solutions to the equation $x^2 - 5x + 1 = 0$ by plotting a graph.
>
>
>
> There are two solutions: $x = 0.2$ or $x = 4.8$
>
> These solutions are approximate.

Draw the graph of $y = x^2 - 5x + 1$ (see p.48).

- This method can also be used to solve equations where the solutions are the points of intersection between a curve and a line.

Using the Quadratic Formula

- The quadratic formula is used to solve equations in the form $ax^2 + bx + c = 0$

 The quadratic formula is $x = \dfrac{-b \pm \sqrt{b^2 - 4ac}}{2a}$

Key Point

Remember, when calculating b^2, a negative number squared is positive.

Using the quadratic formula, solve $4x^2 - x - 2 = 0$

$$x = \dfrac{1 \pm \sqrt{(-1)^2 - 4 \times 4 \times (-2)}}{2 \times 4}$$

$x = 0.843$ or -0.593

Substitute in: $a = 4$, $b = -1$, $c = -2$

Completing the Square

- To **complete the square**, you must write the equation in the form $(x + p)^2 + q = 0$ and then solve it.
- If the **coefficient** of the x^2 term is **not** 1, you must take the coefficient out as a factor and then complete the square.

Solve the equation $x^2 - 2x - 2 = 0$

$x^2 - 2x = 2$

$(x - 1)^2 - 1^2 = 2$

$(x - 1)^2 - 1 = 2$

$(x - 1)^2 = 3$

$x = 1 \pm \sqrt{3}$

Move the constant over to the right-hand side.

The number in the brackets is always half the coefficient of x.

Always minus the square of that number.

Solve.

Key Point

Iteration involves solving an equation many times. With each iteration your answer becomes more accurate.

Iteration

- **Iteration** is the repetition of a mathematical procedure.
- The results of one iteration are used as the starting point for the next.

Solve $x^2 - 3x - 1 = 0$ using iteration, to 2 decimal places.

$x = \dfrac{1}{x - 3}$ or $x = \dfrac{1}{x} + 3$

$x_{n+1} = \dfrac{1}{x_n - 3}$ or $x_{n+1} = \dfrac{1}{x_n} + 3$

The solutions are $x = -0.30$ and $x = 3.30$

Rearrange to make x the subject.

These two equations give the two solutions.

Substitute in any value for x_n. Use the results as a starting point for the next substitution. Continue until the terms converge.

Key Words

factorisation
intersection
complete the square
coefficient
iteration

Quick Test

1. Solve the equation $x^2 = 2x + 5$ by the method of intersection.
2. Solve the equation $3x^2 - 5x - 1 = 0$ using the quadratic formula.
3. Solve the equation $x^2 - 4x + 2 = 0$ by completing the square.
4. Solve $x^2 - 5x - 1 = 0$ using iteration (to 2 d.p.).

Simultaneous Equations and Functions

You must be able to:

- Solve linear and quadratic simultaneous equations algebraically
- Find approximate solutions to simultaneous equations using a graph
- Translate simple situations into two simultaneous equations and solve
- Work out composite functions.

Algebraic Methods

- **Linear simultaneous equations** can be solved by elimination.

Solve the following simultaneous equations:

$$3x - y = 18 \qquad \text{Equation 1}$$
$$x + y = 10 \qquad \text{Equation 2}$$
$$4x = 28 \quad \leftharpoon \quad 7 + y = 10 \leftarrow$$
$$x = 7 \qquad\qquad y = 3$$

> **Key Point**
>
> Solutions to simultaneous equations always come in pairs.

Add equation 1 and equation 2 to eliminate the y terms.

Substitute your value for x into one of the equations.

Annabel buys three pears and two apples for £1.20
David buys four pears and three apples for £1.65

Work out the cost of one apple and one pear.

$$3p + 2a = 120 \qquad \text{Equation 1} \leftarrow$$
$$4p + 3a = 165 \qquad \text{Equation 2}$$
$$\text{Equation 1} \times 4: \ 12p + 8a = 480 \leftarrow$$
$$\text{Equation 2} \times 3: \ 12p + 9a = 495$$
$$a = 15 \leftarrow$$
$$3p + (2 \times 15) = 120 \leftarrow$$
$$p = 30$$

An apple costs 15p and a pear costs 30p.

Form two equations with the information given.

Multiply so that the p terms match. Remember to multiply all terms.

Subtract equation 1 from equation 2.

Substitute your value for a into one of the equations and solve.

- When you have a non-linear equation and a linear equation, always substitute the linear into the non-linear.

Solve the equations $2y + x = 3x^2$ and $y + 5x = -3$

$$2y + x = 3x^2 \qquad \text{Equation 1}$$
$$y + 5x = -3 \qquad \text{Equation 2}$$
$$y = -3 - 5x \leftarrow$$
$$2 \times (-3 - 5x) + x = 3x^2 \leftarrow$$
$$3x^2 + 9x + 6 = 0 \leftarrow$$
$$x^2 + 3x + 2 = 0 \leftarrow$$
$$(x + 1)(x + 2) = 0 \leftarrow$$
$$x = -1, \ y = 2 \ \text{or} \ x = -2, \ y = 7$$

Make y the subject of equation 2.

Substitute into equation 1.

This is a quadratic equation; rearrange to = 0.

Divide all terms by 3.

Solve by factorisation.

Solving Equations with Graphs

- You can plot graphs and find the point of intersection. However, the solutions will often only be approximate.

Solve the equations $y = 3x^2$ and $y + 5x = 3$

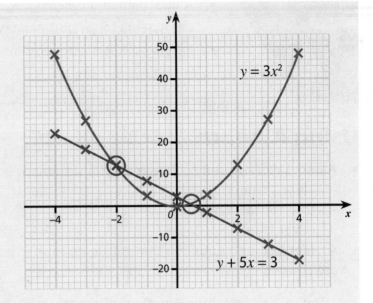

The points of intersection are $(-2.1, 13.5)$ and $(0.5, 1)$.

So, the two approximate solutions are $x = 0.5$, $y = 1$ or $x = -2.1$, $y = 13.5$

Functions

- Functions have an **input** and an **output**, e.g.

  If $f(x) = 4x + 2$ $f(5) = 4 \times 5 + 2 = 22$ ← 5 is the input of the function, so substitute for x.

- The inverse function is the reverse process of the function, e.g.

  If $f(x) = 4x + 2$ $f^{-1}(x) = \dfrac{x-2}{4}$

> **Key Point**
>
> The inverse function is written as $f^{-1}(x)$.

- A **composite function** is when one function becomes the input of another, e.g.

  This means make $g(x)$ the input of $f(x)$.

  $g(x) = 3x - 2$

  Expand the brackets to simplify.

  $f(x) = 2x^2 + 5$

  This means make $f(x)$ the input of $g(x)$.

  a) Work out $fg(x)$ **b)** Work out $gf(x)$

  Expand the brackets to simplify.

  $fg(x) = f(3x - 2)$ $gf(x) = g(2x^2 + 5)$

  $\quad\quad = 2(3x - 2)^2 + 5$ $\quad\quad = 3(2x^2 + 5) - 2$

  $\quad\quad = 18x^2 - 24x + 13$ $\quad\quad = 6x^2 + 13$

1. Solve the simultaneous equations $2x + y = 5$ and $x + y = 3$
2. Solve the simultaneous equations $y = 3x - 1$ and $y = x^2 - 5$
3. Solve the simultaneous equations $y = x$ and $y = x^2 - 2$ using a graph.
4. If $f(x) = 3x + 1$ and $g(x) = 2x^2 + 5$, work out $gf(x)$.

> **Key Words**
>
> linear equation
> simultaneous equations
> function
> input
> output
> composite functions

Algebraic Proof

You must be able to:

- Argue mathematically
- Use algebra to support and construct arguments
- Include proofs in your arguments.

Using Algebra

- Algebra can be used to determine if a statement is:
  - always true
  - true for certain values
  - never true.
- An **identity** is an equation that is true no matter what values are chosen.

Key Point

$\equiv$ means **always** equal to

$\equiv$ is the identity sign and it means that any equation it is used in will always be true for any value of the variable(s).

The sum of any three consecutive integers is always a multiple of 3.

Is this statement true? You must show your working.

Let n represent the first integer.
The second integer is $n + 1$ and the third integer is $n + 2$.
The sum is: $n + (n + 1) + (n + 2) \equiv 3n + 3$
This can be written as $3(n + 1)$
Therefore, the sum of three consecutive integers is always a multiple of 3. The statement is **always** true.

The sum of the squares of any three consecutive integers is always a multiple of 3.

Is this statement true? You must show your working.

Let n represent the first integer.
The second integer is $n + 1$ and the third integer is $n + 2$.
The sum is of the squares is: $n^2 + (n + 1)^2 + (n + 2)^2 \equiv 3n^2 + 6n + 5$
3 is **not** a factor of this expression, therefore the statement is **never** true.

Algebraic Proof

- Algebra can be used to help construct an **argument**.
- **Proofs** are examples of mathematical reasoning, used in an argument, to show a statement is true, i.e. they provide evidence for an argument.

The median of three consecutive integers is n. Show that the mean is also n.

The median is the middle number, so the three consecutive integers in terms of n are: $n - 1$, n, $n + 1$

$$\text{Mean} = \frac{(n - 1) + n + (n + 1)}{3} = \frac{3n}{3} = n$$

Show that $\dfrac{3}{x + 1} + \dfrac{3x}{x^2 + 3x + 2} \equiv \dfrac{6}{x + 2}$

$$\frac{3}{x + 1} + \frac{3x}{x^2 + 3x + 2} \equiv \frac{3}{x + 1} + \frac{3x}{(x + 1)(x + 2)}$$

> Start with the LHS (left-hand side). Factorise the denominator.

$$\frac{3}{x + 1} + \frac{3x}{(x + 1)(x + 2)} \equiv \frac{3(x + 2)}{(x + 1)(x + 2)} + \frac{3x}{(x + 1)(x + 2)}$$

> Find a common denominator.

$$\frac{3(x + 2)}{(x + 1)(x + 2)} + \frac{3x}{(x + 1)(x + 2)} \equiv \frac{3(x + 2) + 3x}{(x + 1)(x + 2)}$$

> Simplify.

$$\frac{3(x + 2) + 3x}{(x + 1)(x + 2)} \equiv \frac{6x + 6}{(x + 1)(x + 2)}$$

$$\frac{6x + 6}{(x + 1)(x + 2)} \equiv \frac{6(x + 1)}{(x + 1)(x + 2)}$$

> Factorise the numerator.

$$\frac{6(x + 1)}{(x + 1)(x + 2)} \equiv \frac{6}{(x + 2)} = \text{RHS}$$

> Cancel down.

The diagram shows two squares. The blue square has a side length of $2n$.

Show that the ratio of the area of the green square to the area of the blue square is $1 : 2$

Key Point

Always show every stage of your working.

Area of blue square is: $2n \times 2n = 4n^2$

The side length of the green square is: $\sqrt{n^2 + n^2} = \sqrt{2n^2}$

> Use Pythagoras' Theorem.

The area of the green square is: $\sqrt{2n^2} \times \sqrt{2n^2} = 2n^2$

Ratio of areas is: $2n^2 : 4n^2$, which simplifies to $1 : 2$

Quick Test

1. Show that the sum of any two consecutive integers will always be an odd number.
2. Show that the difference of the squares of two consecutive integers will always be an odd number.
3. A set of five consecutive integers has a median of n. Work out the mean in terms of n.
4. Show that $\dfrac{2}{x} + \dfrac{3}{x^2} \equiv \dfrac{2x + 3}{x^2}$

Key Words

identity
argument
proof

Circles

You must be able to:

- Identify and use different properties of circles
- Apply and prove theorems relating to circles
- Understand cyclic quadrilaterals.

Circle Theorems

- The angle **subtended** (formed) at the centre of a circle is twice the angle at the **circumference** if both are subtended from the same **chord** or **arc**.

- Angles subtended from the same chord or arc to the circumference are equal.

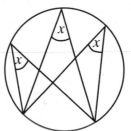

- Any angle subtended from each end of the diameter will be 90°.

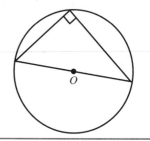

Cyclic Quadrilaterals

- All four corners of a **cyclic quadrilateral** touch the circumference of the circle.
- Opposite angles of a cyclic quadrilateral add up to 180°.

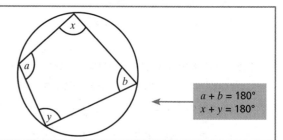

$a + b = 180°$
$x + y = 180°$

Tangents and Chords

- A **tangent** is a straight line that touches the circumference of the circle at a single point.
- If a tangent and a **radius** meet at a point on the circumference, the angle between them is 90°.

- If two tangents are drawn from the same external point to a circle, they will be of equal lengths.
  $AB = AC$
  Triangle OAB and triangle OAC are congruent.

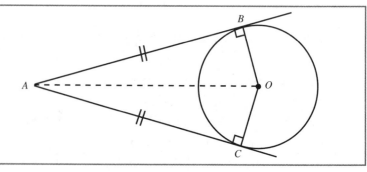

- The line joining the centre of the circle to the midpoint of a chord is perpendicular to the chord.

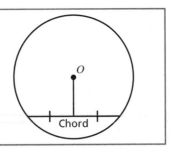

Sectors

- The **sector** of a circle is a section of the circle bounded by two radii and an arc.
- If the angle at the centre is θ then:

Sector Area $= \dfrac{\theta}{360} \times \pi r^2$

Arc Length of Sector $= \dfrac{\theta}{360} \times 2\pi r$

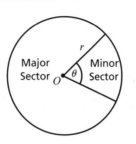

Alternate Segment Theorem

- The angle between the tangent and the chord is equal to the angle subtended by the chord in the alternate **segment**.

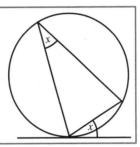

Quick Test

1. Work out the size of the lettered angle in each diagram.

a)

b)

c)

d)

Vectors

You must be able to:

- Add and subtract vectors
- Multiply a vector by a scalar
- Work out the magnitude of a vector
- Use vectors in geometric arguments and proofs.

Properties of Vectors

- A **vector** is a quantity that has both **magnitude** (size) and direction.
- Vectors are equal only when they have equal magnitudes and are in the same direction.

> **Key Point**
>
> The direction of a vector is shown by an arrow.

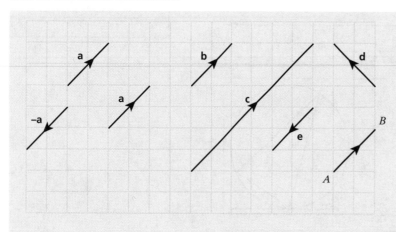

b = **a** (same direction, same length)

d ≠ **a** (different direction, same length)

e = –**a** (opposite direction, same length)

c = 3**a** (same direction, 3 × length of **a**)

$$\mathbf{a} = \overrightarrow{AB} = \underline{\mathbf{a}} = \begin{pmatrix} 2 \\ 2 \end{pmatrix}$$

These are all ways of writing the same vector.

$$-\mathbf{a} = \begin{pmatrix} -2 \\ -2 \end{pmatrix}$$

a and **c** are parallel vectors.

a and **b** are equal vectors.

- Any number of vectors can be added together.

$$\mathbf{a} + \mathbf{b} + \mathbf{c} + \mathbf{d} = \begin{pmatrix} 3 \\ 6 \end{pmatrix} + \begin{pmatrix} 2 \\ -2 \end{pmatrix} + \begin{pmatrix} 6 \\ 3 \end{pmatrix} + \begin{pmatrix} 3 \\ -6 \end{pmatrix}$$

$$= \begin{pmatrix} 14 \\ 1 \end{pmatrix}$$

$$\overrightarrow{AB} + \overrightarrow{BC} + \overrightarrow{CD} + \overrightarrow{DE} = \overrightarrow{AE} \text{ or } \mathbf{e}$$

This is the resultant vector.

- When a vector is multiplied by a **scalar** (a numerical value), the resultant vector will always be parallel to the original vector.
- When a vector is multiplied by a positive number (not 1), the direction of the vector does not change, only its magnitude.

> **Key Point**
>
> The sum of the lengths $AB + BC + CD + DE$ does **not** equal the length of AE.

- When a vector is multiplied by a negative number (not –1), the magnitude of the vector changes and the vector points in the opposite direction.
- The magnitude of a vector **a** is written $|a|$
- Magnitude of vector $\begin{pmatrix} x \\ y \end{pmatrix}$ is $\sqrt{x^2 + y^2}$

Work out the magnitude of vector **a**.

Vector $\mathbf{a} = \begin{pmatrix} 5 \\ -8 \end{pmatrix}$

$a^2 = 5^2 + 8^2$ ← Use Pythagoras' Theorem.

$a^2 = 25 + 64$

$|a| = \sqrt{89}$

$|a| = 9.43$ (to 3 significant figures)

Vectors in Geometry

- All rules of algebra can be applied to vector expressions.

In triangle DEF, G and H are the midpoints of DE and DF.

Prove that $GH = \frac{1}{2}EF$ and that GH is parallel to EF.

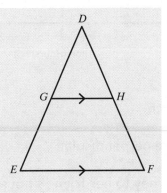

Let $\vec{DE} = \mathbf{e}$ and $\vec{DF} = \mathbf{f}$

So, $\vec{EF} = \vec{ED} + \vec{DF}$

$= -\mathbf{e} + \mathbf{f} = \mathbf{f} - \mathbf{e}$

Also $\vec{DG} = \frac{1}{2}\vec{DE} = \frac{1}{2}\mathbf{e}$

$\vec{DH} = \frac{1}{2}\vec{DF} = \frac{1}{2}\mathbf{f}$

So, $\vec{GH} = \vec{GD} + \vec{DH}$

$= -\frac{1}{2}\mathbf{e} + \frac{1}{2}\mathbf{f} = \frac{1}{2}\mathbf{f} - \frac{1}{2}\mathbf{e} = \frac{1}{2}(\mathbf{f} - \mathbf{e})$

$\vec{GH} = \frac{1}{2}\vec{EF}$

Therefore, $GH = \frac{1}{2}EF$ and GH is parallel to EF. ← This is called the midpoint theorem.

Quick Test

1. Vector **a** has a magnitude of 3cm and a direction of 120°. Vector **b** has a magnitude of 4cm and a direction of 040°. Draw the vectors:
   a) a   b) b   c) –a   d) 2a   e) a + b

Key Words

vector
magnitude
scalar

Review Questions

Uses of Graphs & Other Graphs 1 & 2

1 The graph below shows the journey of a car.

Estimate the total distance travelled.

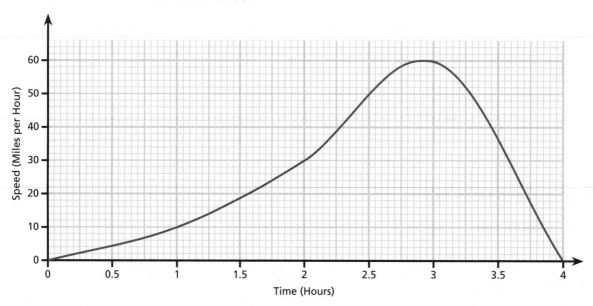

Time (Hours)

[3]

2 Line 1 has equation $5y = 3x - 2$
Line 2 has equation $3y + 5x = 6$

Show that line 1 and line 2 are perpendicular. [2]

3 The graph shows Sophie's journey to her friend's house. Her friend lives 18km away.
Sophie begins her journey at 1pm.

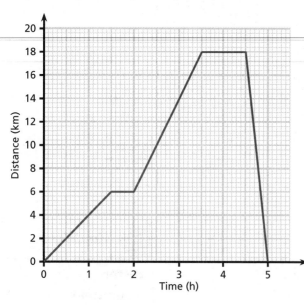

a) Sophie stops on the way to see her friend.

    i) How is this shown on the graph? [1]

    ii) How long does Sophie stop for? [1]

b) At what time does Sophie arrive at her friend's house? [1]

c) Sophie is picked up from her friend's house by her mum.

    i) Calculate the average speed, in km/h, of her journey home. [2]

    ii) At what time does Sophie arrive home? [1]

Uses of Graphs & Other Graphs 1 & 2 (Cont.)

4 The graph below has equation $y = x^2$

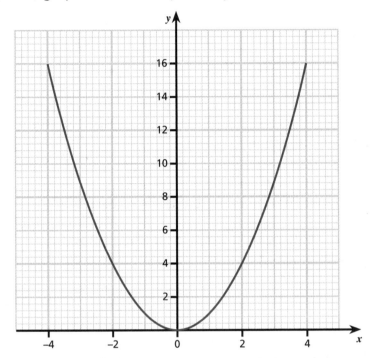

Estimate the gradient of the curve at the point (2, 4). [1]

5 A circle has centre (0, 0) and goes through the point (3, 5).

a) Work out the equation of the circle. [3]

b) Work out the equation of the tangent to the circle at the point (3, 5). [3]

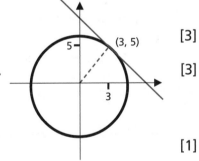

6 **a)** y is directly proportional to x.

Sketch a graph to show this relationship. [1]

b) y is inversely proportional to x.

Sketch a graph to show this relationship. [1]

7 **a)** Expand $(x - 3)(x + 3)(x - 2)$ [3]

b) Sketch the graph of $(x - 3)(x + 3)$ [2]

8 Work out the equation of the line that is perpendicular to the line with equation $2y + x = 5$ and goes through the point (3, 6). [3]

Total Marks .... / 28

Inequalities

1 Work out all the possible integer values of n if $3 \leqslant n \leqslant 7$. [1]

2 Solve $2x - 6 > 2$ [2]

3 Work out all the possible integer values for y if $12 \leqslant 3y \leqslant 36$. [2]

4 Solve $x^2 + 5x - 24 \geqslant 0$ [3]

5 Solve $2 - 9x > 3$ and show the solution on a number line. [2]

6 A TV salesperson is set a target to sell more than six televisions a week.
The manufacturer can let the salesperson have a maximum of 20 televisions each week.

Use an inequality equation to represent the number of televisions that could be sold
each week if the salesperson meets or exceeds their target. [2]

> **Total Marks** .... / 12

Congruence and Geometrical Problems

1 A triangle has angles of 56°, 64° and 60°. The triangle is enlarged by scale factor 2.

What are the angles of the enlarged triangle? [3]

2 In triangle ABC, $AB = 10$cm, $BC = 12$cm and $CA = 10$cm.

a) What type of triangle is triangle ABC? [1]

b) D is a point on BC such that AD is the perpendicular bisector of BC.

Prove that triangle ABD is congruent to triangle ACD. [3]

c) Calculate the area of triangle ABC. [3]

> **Total Marks** .... / 10

Right-Angled Triangles

1 One of the pyramids of Egypt is built on a square base of width 200m.

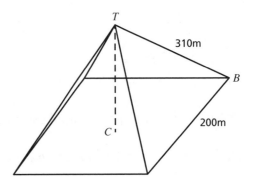

C is the centre of the base. The slant height *TB* is 310m.

Work out the perpendicular height of the pyramid to the nearest metre. **[4]**

2 A rectangle measures 3cm by 4cm.

What is the length of its diagonal? **[1]**

3 Calculate the area of the triangle. **[3]**

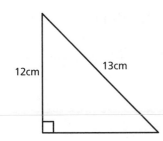

4 *PCDEF* is a pyramid with a rectangular base *CDEF*.
CD = 8cm and *DE* = 5cm.
P is vertically above the centre of the rectangle and *PC* = *PD* = *PE* = *PF* = 13cm.

Calculate:

a) The angle between *PC* and the plane *CDEF*. Give your answer to 1 decimal place. **[4]**

b) The vertical height of *P* above the rectangular base. Give your answer to 3 significant figures. **[2]**

5 Two polar bears, Snowy and Blizzard, are asleep in an enclosure.
The distance from Snowy to Blizzard is 25m. The bearing of Blizzard from Snowy is 078°.

Calculate how far east Blizzard is from Snowy. Give your answer to 2 decimal places. **[2]**

Total Marks .... / 16

Sine and Cosine Rules

1 Here is a triangle ABC.

a) Which angle is the smallest?
Give a reason for your answer. [2]

b) Calculate the size of the smallest angle to the nearest degree. [4]

c) Calculate the area of triangle ABC. [3]

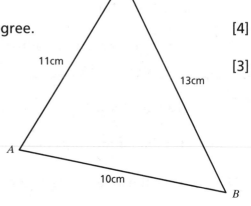

2 A fishing boat leaves harbour and sails 12km on a bearing of 040° to a lighthouse.

The boat then changes direction and sails on a bearing of 145° to visit Seal Island, which is 14km due east from the harbour.

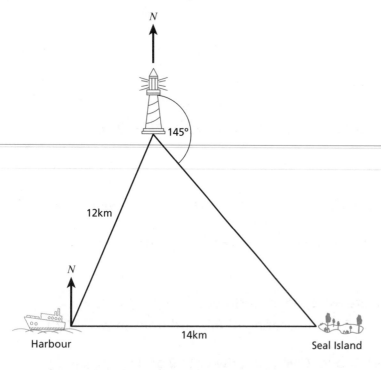

How far is Seal Island from the lighthouse?
Give your answer to 1 decimal place. [4]

Total Marks _____ / 13

Statistics 1 & 2

1 This histogram gives information about the books sold in a bookshop one Friday.

a) Use the histogram to complete the table. [2]

Price (pounds)	Frequency
$0 < P \leqslant 5$	
$5 < P \leqslant 10$	
$10 < P \leqslant 20$	
$20 < P \leqslant 40$	

b) Estimate the mean price of books sold. [3]

c) Work out the class interval that contains the median. [2]

2 Corinna wants to sample 100 people in Malmesbury to find out how often people visit the library. She stands by the entrance of the library on a Thursday morning and asks the first 100 people she meets.

a) Explain what is wrong with this sample. [1]

b) Describe a better method that Corinna could use to collect the data. [2]

3 The scatter diagram below shows the body temperature and pulse rate of 10 animals.

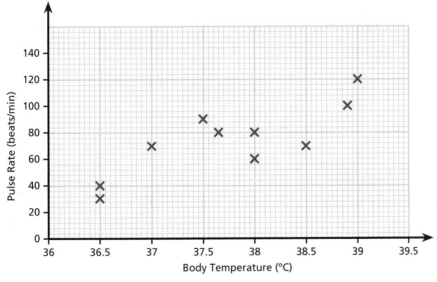

a) Draw a line of best fit and use it to estimate the pulse rate of an animal with a body temperature of 39.5°C. [2]

b) Comment on the reliability of this estimate. [1]

Total Marks .... / 13

Practice Questions

Measures, Accuracy and Finance

1 Estimate an answer to the following calculation: 📱

$$\frac{512 \times 7.89}{0.19}$$

[2]

2 4kg of carrots and 5kg of potatoes cost £6.11

If 5kg of carrots cost £4.45, work out the cost of 4kg of potatoes. [3]

3 Sabrina's take-home pay is worked out using this formula:

Take-Home Pay = Hours Worked × Hourly Rate – Deductions

Sabrina's hourly rate is £35.
Her deductions were £218.
Her take-home pay was £657.

Work out the number of hours she worked that week. [3]

4 Calculate $\dfrac{\sqrt{(6.2^2 - 3.6)}}{2.6 \times 0.15}$

Give your answer to **a)** 2 decimal places and **b)** 3 significant figures. [4]

5 The price of a new television is £560, including tax at 17.5%.

Work out the cost of the television before tax is added.
Give your answer to the nearest penny. [2]

6 Graham uses the following formula to work out his gas bill:

Cost = Standing Charge + Cost per Unit × Units Used

The standing charge is £50.93 and the cost per unit is 3p:

a) Last month Graham's gas bill was £53.15. How many units did he use? [2]

b) The standing charge is increased by 3%. What is the new standing charge? [2]

Total Marks _____ / 18

Solving Quadratic Equations

1 Solve the equation $3x^2 = 27$ [1]

2 **a)** Write $x^2 + 8x - 12$ in the form $(x + p)^2 + q$, where p and q are integers. [2]

b) Solve the equation $x^2 + 8x - 12 = 0$ giving your answer in the form $a \pm b\sqrt{7}$. [2]

c) Write down the minimum point of $y = x^2 + 8x - 12$ [1]

3 The diagram shows a trapezium. All the measurements are in centimetres. The area of the trapezium is 14cm².

a) Show that $6x^2 - 5x - 14 = 0$ [3]

b) Solve the equation to find the value of x. [2]

$4x - 6$

$2x$

$2x + 1$

Total Marks .... / 11

Simultaneous Equations and Functions

1 **a)** On the same set of axes, draw the graphs $y = 2x^2$ and $y = 3x + 2$ [2]

b) Use your graph to solve the equation $2x^2 = 3x + 2$ [2]

2 Solve the simultaneous equations: [4]

$2x + y = 1$

$y = x^2 - 2$

3 $f(x) = x^2 + 1$ and $g(x) = 3x - 2$

a) Work out $fg(x)$ [2]

b) $fg(x) = gf(x)$

Show that this statement is incorrect. [2]

c) Use the quadratic formula to solve the equation $fg(x) = 10$ [3]

Total Marks .... **Practise** / 15

Algebraic Proof

1 Prove that the sum of the squares of any three consecutive integers is 1 less than a multiple of 3. [3]

2 Grace surprises her friends with the following number trick:
- Think of a number.
- Multiply your number by 3.
- Add 30 to your answer.
- Divide your number by 3.
- Subtract the number you first thought of.
- Your answer is 10.

Use algebra to prove that Grace's trick always works. [3]

3 Show that $\dfrac{3}{x+2} + \dfrac{9}{x^2+x-2} = \dfrac{3}{x-1}$ [3]

4 Show that $\dfrac{6}{3x+2} + \dfrac{4}{3x-2} \equiv \dfrac{30x-4}{9x^2-4}$ [3]

5 The diagram below shows a circle inside a square. The square has a side length of $6x$.

Show that the ratio of the area of the square to the area of the circle is $4 : \pi$ [4]

6 Prove that $(4n+1)^2 - (4n-1)^2$ is a multiple of 8 for all positive integer values of n. [3]

7 Prove that the difference between the squares of any two consecutive even numbers is always a multiple of 4. [3]

8 a and b are numbers such that $a = b + 2$

The sum of a and b is equal to the product of a and b.

Show that a and b are **not** integers. [3]

Total Marks _____ / 25

Circles

1 For each of the following questions, work out the lettered angles.
The centre of each circle (where appropriate) is marked with an *O*. 🖩

a) [1]

b) [2]

c) [2]

d) [2]

e) [2]

f) [1]

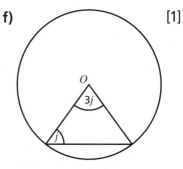

Total Marks .... / 10

Vectors

1 *ABC* is a triangle. *D* is the midpoint of *AB*. *E* is the midpoint of *BC*.
$\overrightarrow{AB}$ = **a** and $\overrightarrow{AC}$ = **c**

  a) Work out $\overrightarrow{AE}$ in terms of **a** and **c**. [3]

  b) Show that $\overrightarrow{DE}$ is parallel to $\overrightarrow{AC}$. [3]

2 State whether the following can best be described as a vector or as a scalar:

  a) The length of a desk [1]

  b) The flight path of an aircraft [1]

  c) A snooker ball hit directly into a pocket. [1]

Total Marks .... / 9

Measures, Accuracy and Finance

1 Calculate the value of $\dfrac{25.75 \times 31.3}{7.62 - 1.48}$

Give your answer to **a)** 2 decimal places and **b)** 3 significant figures. [3]

2 The Cotton family, comprising of Mum, Dad and 16-year-old daughter, are going away to Turkey.

The brochure states that for every three days booked, you receive one day free. The price per adult per day is £42. There is a 15% reduction for a child up to the age of 17. The package is all inclusive.

a) What would the total cost of the holiday be if the family books 12 days? [3]

b) There is an additional early booking discount of 5%.

If the Cotton family book early, how much do they save? [2]

c) The flight to Turkey takes 4 hours 20 minutes. The departure time from Gatwick is scheduled for 11.10am but, due to bad weather, the flight is delayed by 1 hour 34 minutes.

If Turkish time is two hours ahead of UK time, at what time did the flight arrive in Turkey? [2]

3 What is the maximum area of a square of side 4.2cm measured to the nearest millimetre? [1]

4 Meka wants to buy a mobile phone. There are two different options:

Option 1: Phone costs £146 + £80 per month

Option 2: Phone costs £548 + £43 per month

Both options come with a 5% annual discount.

a) Which option is cheaper for a one-year period? [3]

b) How much cheaper is this option? [1]

Total Marks .... / 15

Solving Quadratic Equations

1 A circle has equation $x^2 + y^2 = 36$

    a) Write down the equation of the tangent shown
on the diagram. [1]

    b) A chord, as shown on the diagram, goes through
the point (0, 3) and intersects the circle in two places.
Work out the points of intersection. [3]

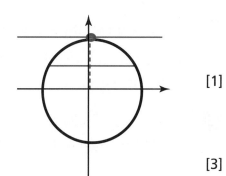

2 **a)** Write $x^2 - 12x + 26$ in the form $(x + p)^2 + q$. [3]

    b) Use your answer from part **a)** to solve the equation $x^2 - 12x + 26 = 0$
Leave your answers in surd form. [2]

3 Solve the equation $\dfrac{4}{x} + \dfrac{4}{x + 2} = 3$ [4]

4 The area of the triangle is 4cm².

    Work out the value of x. Give your answer in the form $a\sqrt{6}$. [3]

$\frac{x}{15}$ cm

$\frac{x}{5}$ cm

5 Use the quadratic formula to solve the equation $x^2 - 7x - 15 = 0$ [2]

> **Total Marks** .... / 18

Simultaneous Equations and Functions

1 **a)** On the same axes, draw the graphs of $y = 3x^2$ and $y = 4x + 2$. [2]

    b) Use your graphs to find estimates to the solutions of the equation $3x^2 = 4x + 2$ [2]

2 Solve the following simultaneous equations:

$$y - x = 4$$

$$y = x^2 + 2$$ [4]

> **Total Marks** .... / 8

Algebraic Proof

1 **a)** Write down the expression for the nth term of the sequence 4, 7, 10, 13, 16 ... [1]

  b) Prove that the product of any two consecutive terms in this sequence is also a term
  of the sequence. [3]

2 Prove algebraically that the difference between the squares of any two consecutive
integers is equal to the sum of these two integers. [3]

3 Show that $\dfrac{x^2 + 5x + 6}{x^2 + 2x} = \dfrac{x + 3}{x}$ [3]

4 Jessa thinks that when you subtract 7 from a number, the answer is always odd.

Give an example to show that Jessa is wrong. [1]

5 The shape shown is a square.

Emma states that the area of the square is half of the product of the diagonals.
Prove that Emma is correct. [2]

6 Prove that $(3n + 1)^2 - (3n - 1)^2$ is a multiple of 4 for all positive integer values of n. [3]

7 Here is a sequence of numbers:
2, 5, 10, 17, 26, 37 ...

  a) Write down the expression for the nth term of this sequence. [1]

  b) Show algebraically that the sum of two consecutive terms in this sequence
  will always be odd. [3]

8 Show that $\dfrac{2}{x - 2} - \dfrac{8}{x^2 - 4} \equiv \dfrac{2}{x + 2}$ [3]

9 Prove that the square of any odd number is always one more than a multiple of 4. [3]

10 *'The sum of two prime numbers is also a prime number.'*

Give an example to disprove this statement. [1]

Total Marks .... / 27

Circles

1 Work out the size of the angle marked with a letter. The centre of each circle (where appropriate) is marked with an O. 🖩

a) [1]

b) [1]

c) [2]

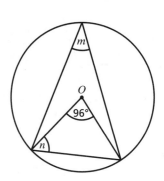

2 a) Label each arrow pointing to a different part of the circle. [5]

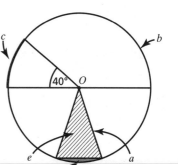

b) Calculate the length of c if the radius = 10cm (take π = 3.14).
Give your answer to 2 decimal places. [3]

Total Marks _____ / 12

Vectors

1 $ABCD$ is a parallelogram and AC is one of the diagonals.
P is a point on AC such that $AP = \frac{1}{4} AC$
$\overrightarrow{DA} = 4\mathbf{a}$ and $\overrightarrow{DC} = 4\mathbf{c}$

a) Work out the vector $\overrightarrow{AC}$ in terms of $\mathbf{a}$ and $\mathbf{c}$. [1]

b) Work out the vector $\overrightarrow{AP}$ in terms of $\mathbf{a}$ and $\mathbf{c}$. [1]

c) Work out the vector $\overrightarrow{DP}$ in terms of $\mathbf{a}$ and $\mathbf{c}$. [1]

2 Write down whether each of these statements is **true** or **false**.

a) Velocity is a vector. [1]

b) Two vectors are equal if they have the same magnitude but are in opposite directions. [1]

Total Marks _____ / 5

1 The diagram shows a regular pentagon of side length P.

Shade the area on the diagram that is represented by the expression $\frac{1}{2}CG + \frac{1}{2}PG$.

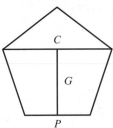

[1]

2 Factorise completely $3x^3 + 6x^2$

Answer _____ [1]

3 Solve $3(x + 3) - 5 = 2(x - 2)$

Answer _____ [3]

4 Work out the value of $\frac{5}{7}\left(x - \frac{2}{5}\right) + 9y$ when $x = \frac{1}{5}$ and $y = \frac{2}{7}$

Answer _____ [2]

5 Simplify $3x^2 + 6x - 2x^2 + 4x - 2y - 6x^2$

Answer _____ [2]

6 Simplify $3p^2y - py + p^2y + 7py$

Answer _____ [2]

7 Solve $5(x - 3) = -3$

Answer _____ [2]

8 Write $(2x - 2)(x - 3)(3x + 4)$ in the form $ax^3 + bx^2 + cx + d$ where a, b, c and d are integers.

Answer _____ [2]

9 **a)** Factorise $x^2 - x - 2$

Answer _____ [1]

b) Factorise $2x^2 - 5x + 2$

Answer _____ [1]

c) Using your answers from parts **a)** and **b)** simplify $\dfrac{2x^2 - 5x + 2}{x^2 - x - 2}$

Answer _____ [2]

10 Use the formula $P = 3r - q^2$ to work out the exact value of q when $P = 30$ and $r = 20$.

Answer _____ [2]

11 **a)** Factorise $x^2 - 16$

Answer _____ [1]

b) Solve $x^2 - 16 = 0$

Answer _____ [2]

12 The formula used to calculate the area of a circle is $A = \pi r^2$.

A circle has an area of 25cm^2.

Ethan thinks the radius of the circle is $\dfrac{5}{\sqrt{\pi}}$

Guy thinks the radius is $\dfrac{\sqrt{\pi}}{5}$

Who is correct? You must show how you got your answer.

_____

_____ [2]

13 Expand and simplify $\sqrt{5}\left(\sqrt{5} + 3\right)$

Answer _____ [2]

14 Write the following numbers in order, smallest first.

| 6.77 | 6.767 | 6.677 | 6.8 |

Answer _____ [1]

15 **a)** Write 45 as a product of prime factors.

Answer _____ [2]

b) Write 105 as a product of prime factors.

Answer _____ [1]

c) Use your answers to parts **a)** and **b)** to work out the highest common factor (HCF) of 45 and 105.

Answer _____ [2]

d) Work out the lowest common multiple (LCM) of x^2y and y^2x, where x and y are prime numbers.

Answer _____ [2]

16 Work out $5\frac{1}{6} - 2\frac{1}{3}$

Answer _____ [3]

17 $P = xy$

x is increased by 10%.

y is increased by 10%.

Work out the percentage increase in P.

Answer _____ [2]

18 Mandeep is looking for a new 12-month phone contract.

Dave's Dongles
£12 a month
+
5p a minute
10% discount on first 6 months

Ian's Internet
£10 a month
+
6p a minute
15% discount on first 4 months

On average Mandeep uses 120 minutes per month.

Which phone contract is cheaper for Mandeep?
You must show your working.

Answer _____ [5]

19 97 × 1452 = 140 844

Use this information to write down the value of:

a) 9.7 × 145.2

Answer _____ [1]

b) 0.97 × 0.1452

Answer _____ [1]

20 $y = \dfrac{ab}{a + b}$

$a = 3 \times 10^4$

$b = 5 \times 10^3$

Work out the value of y correct to five significant figures.

Give your answer in standard form.

Answer _____ [2]

21 Write down the formula for calculating speed. 🖩

Answer _____ [1]

22 A regular polygon has 20 sides.

Work out the size of each interior angle.

Answer _____ [2]

23 Write down the nth term of the following sequence of numbers:

8, 11, 14, 17, 20 …

Answer _____ [1]

24 A cat rescue centre recorded the age of the cats it rehomed over the course of a year.

Age	Frequency
$0 \leqslant a < 2$	256
$2 \leqslant a < 4$	219
$4 \leqslant a < 6$	165
$6 \leqslant a < 8$	120
$8 \leqslant a < 10$	65

a) Write down the group that contains the median.

Answer _____ [1]

b) Calculate an estimate of the mean age.

Answer _____ [3]

Mixed Exam-Style Questions

c) On the graph paper below draw a histogram to represent the data.

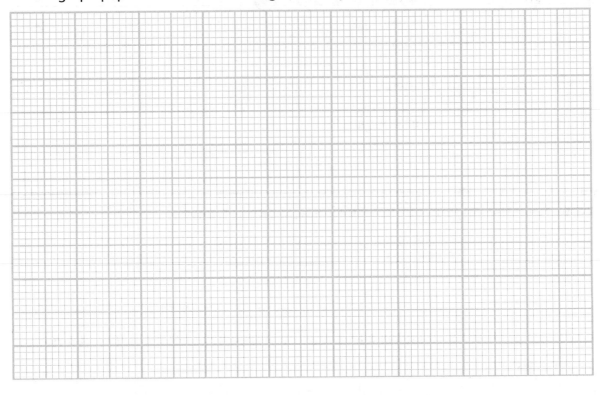

[2]

d) The rescue centre runs a campaign to help rehome the older cats and a further 78 cats over the age of 5 are rehomed.

If this information is added to the data, does it affect the group in which the median lies? Explain your answer.

_____

_____

_____ [2]

25 The force (F) between two magnets is inversely proportional to the square of the distance (x) between them.

$F = 4$ when $x = 3$

a) Work out an expression for F in terms of x.

Answer _____ [2]

b) Work out the value of F when $x = 5$.

Answer _____ [1]

c) Work out the value of x when $F = 20$.

Answer _____ [2]

26 **a)** Work out the length of side x.

6cm

x

9cm

37°

Answer _____ [3]

b) Work out the area of the triangle.

Answer _____ [2]

c) The lengths of the triangle are enlarged by scale factor 3.
The area of the new triangle can be found using the following rule:
Area of Enlarged Triangle = P × Area of the Original

Write down the value of P.

Answer _____ [1]

27 What is the equation of the y-axis?

Answer _____ [1]

Mixed Exam-Style Questions

28 As part of a health and safety review, a company surveys its employees to find out how many wear glasses or contact lenses.

	Male	Female
Glasses	9	6
Contact Lenses	8	16
Neither	20	15

a) Write down the ratio of the number of females who wear glasses to the number of females who wear contact lenses. Give your answer in its simplest form.

Answer _____ [1]

b) What percentage of all employees are male and do not wear glasses or contact lenses?

Answer _____ [2]

29 a) Sketch the graph of
$y = x^2 + 5x + 4$ [1]

b) Using your graph from part **a)**, solve the inequality
$x^2 + 5x \leqslant -4$

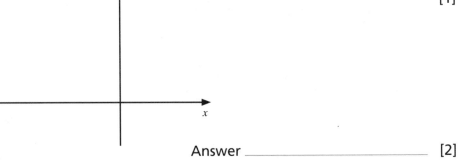

Answer _____ [2]

c) On your graph, shade the region that represents the inequality in part **b)**. [1]

30 The diagram shows a regular hexagon $ABCDEF$ with centre O.

a) $\overrightarrow{OA} = 2\mathbf{a}$ and $\overrightarrow{OB} = 2\mathbf{b}$

Express in terms of $\mathbf{a}$ and/or $\mathbf{b}$

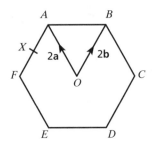

i) $\overrightarrow{AB}$

Answer _____ [1]

ii) $\overrightarrow{EF}$

Answer _____ [1]

b) X is the midpoint of AF.

Express $\overrightarrow{DX}$ in terms of **a** and **b**.

Answer _____ [2]

c) Y is the point on BA extended, such that $BA : AY = 3 : 2$

Prove that D, X and Y lie on the same straight line.

Answer _____ [3]

31 Solve the simultaneous equations:

$y = x^2 - 1$

$y = 3x + 3$

Answer _____ [4]

32 In the diagram, A, B and C are points on the circumference of a circle, centre O.
Angle $BCE = 57°$

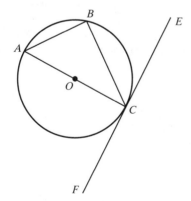

FE is a tangent to the circle at point C.

a) Calculate the size of angle ACB. Give a reason for your answer.

_____ [2]

b) Calculate the size of angle BAC.
Give reasons for your answer.

_____ [2]

33 Simplify $(5t^{-2})^{-1}$

Answer _____ [2]

34 The diagram shows a sector of a circle, centre O.
The radius of the circle is 6cm.
The angle of the sector at the centre of the circle is 115°.

Work out the perimeter of the sector.

Answer _____ [4]

35 A bag contains six blue beads and five red beads.
Samuel takes a bead at random from the bag. He records its colour and replaces it.
He does this one more time.

Work out the probability that he takes one bead of each colour from the bag.

Answer _____ [3]

36 $P = 1.5$ rounded to 1 decimal place.

$Q = 2.65$ rounded to 2 decimal places.

a) Work out the maximum value of PQ.

Answer _____ [2]

b) Work out the minimum value of $\dfrac{P}{Q}$

Answer _____ [2]

37 A sketch of the graph $y = x^2 + 3x + 2$ is shown below.

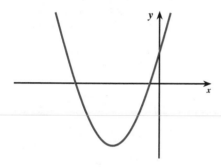

a) Work out the coordinates of the roots of the graph.

Answer _____ [2]

b) Work out the coordinates of the turning point.

Answer _____ [3]

c) Work out the coordinates of the turning point of the curve $y = f(x + 2)$

Answer _____ [1]

Total Marks _____ / 115

Answers

Pages 6–7 Review Questions

1. 2 000 005 **[1]**
2. 45kg **[1]**
3. $2^2 \times 3 \times 5$ **[1]**
4. a) 18 **[1]**
   b) 180 **[1]**
5. 8 **[1]**
6. −8 **[1]**
7. 64 **[1]**
8. $3 + 2m$ **[1]**
9. $3y^2$ **[1]**
10. a) 17, 20 **[1]**
    b) Yes **[1]**; nth term = $3n + 2$ and
    $(3 \times 46) + 2 = 140$ **[1]**
    OR $\frac{(140 - 5)}{3} = 45$
    (term-to-term rule is +3) **[1]**
11. 0.06 **[1]**
12. 9 sides **[1]**
13. 3.46 **[1]**
14. (4.5, 7) **[2]**
15. 27 **[1]**
16. 10km/h **[1]**
17. 125cm³ **[1]**
18. 0.4 **[1]**
19. 0.875 **[1]**
20. $\frac{78}{100}$ **[1]**; $\frac{39}{50}$ **[1]**
21. $\frac{2}{3} = \frac{8}{12} = \frac{16}{24}$ and $\frac{3}{4} = \frac{9}{12} = \frac{18}{24}$ **[1]**; $\frac{17}{24}$ **[1]**
22. 10, 15, 20 **[3]**
23. 999 999 **[1]**
24. £21 : £35 **[2]**
25. 0.04, 0.39, 0.394, 0.4 **[1]**
26. a) $m = 4$ **[1]**
    b) $m = 48$ **[1]**
    c) $m = 1.5$ **[1]**
27. 36cm² **[1]**
28. 404 **[1]**
29. £12.75 **[1]**
30. $13a - 10y$ **[1]**
31. £81 **[1]**
32. $A = \pi \times r \times r$ **[1]**; $A = 314$cm² **[1]**
33. Vol (1) = $2 \times 2 \times 2 = 8$cm³ **[1]**;
    Vol (2) = $4 \times 4 \times 4 = 64$cm³ **[1]**;
    $\frac{64}{8} = 8$ **[1]**
34. 15 **[1]**
35. $\frac{63}{3} \times 4 = 84$ **[2]**
36. $m = -3$ **[1]**
37. (5, 0) **[2]**
38. 13cm **[1]**
39. $(2 \times 3^2) + (3 \times 4) = 18 + 12$ **[1]**; = 30 **[1]**
40. 6.5 **[1]**

Pages 8–31 Revise Questions

Page 9 Quick Test
1. -0.125, 0.0125, $\frac{1}{25}$, 1.25
2. a) 24
   b) 1
   c) −12.5
   d) 1
3. a) 1.63×10^{-3}
   b) 1.63×10^7

Page 11 Quick Test
1. $2^2 \times 3^2$

2. a) 3
   b) 60
3. $m = 3$; $p = 11$
4. 10am

Page 13 Quick Test
1. $6y - 5$
2. −171
3. $12t^2 - 3t$
4. $2r^2(2r - 1)$
5. $t = -0.25$ OR $-\frac{1}{4}$

Page 15 Quick Test
1. $w = 9$
2. $y^2 + 2y - 8$
3. $(2q + 1)(q + 3)$
4. $y = \dfrac{x + 6}{5}$

Page 17 Quick Test
1. £108
2. 4 : 1
3. a) Box A, 1 sachet costs 31.7p; Box B, 1 sachet costs 31.1p; Box B is the best buy
   b) 2

Page 19 Quick Test
1. 402mph
2. £51.11
3. 21%

Page 21 Quick Test
1. a) Rectangle, Parallelogram, Kite
   b) Square, Rhombus
2. Angle $EJH = 16°$

Page 23 Quick Test
1. a) 3240°
   b) 162°
2. 12
3.

Bearing = 150° ± 2°

Page 25 Quick Test
1. $\frac{3}{4} \div \frac{1}{5}$ $\left(\frac{3}{4} + \frac{1}{5} = \frac{15}{20} + \frac{4}{20} = \frac{19}{20}\right.$ and
   $\frac{3}{4} \div \frac{1}{5} = \frac{15}{4} = 3\frac{3}{4}\right)$
2. a) $0.\dot{1}$
   b) recurring
3. $\frac{3}{80}$

Page 27 Quick Test
1. 275 carrots
2. £10.50 + £60 = £70.50
3. 1.5%

Page 29 Quick Test
1. a) 0.1
   b) 45

Page 31 Quick Test
1. $\frac{1}{18}$
2. $\frac{5}{18}$

$$\frac{3}{9} \times \frac{2}{8} + \frac{4}{9} \times \frac{3}{8} + \frac{2}{9} \times \frac{1}{8}$$

Pages 32–37 Practice Questions

Page 32
1. -3.38, -3, $-\frac{1}{8}$, 3.3 **[2]** (1 mark if one mistake; 0 marks if more than one mistake)
2. $-4 \times -4 \times -4 = -64$ **[1]**;
   $(2^3 - 3^3) - (-7)^2 = -19 - 49 = -68$ **[1]**;
   $-4 \times -4 \times -4$ is larger **[1]**
3. a) 3.84×10^5 **[2]** (1 mark for correct digits but decimal point in wrong position)
   b) $384\,000 - 19\,600$ **[1]**; = 364 400 OR 3.644×10^5 **[1]**

1. 84 **[1]**
2. a) 60 **[1]**
   a) 3 **[1]**
3. Fifth square number = 25, third cube number = 27 **[1]**; The third cube number is greater **[1]**
4. $256 = 2^8$ **[1]**; $n = 8$ **[1]**
5. No **[1]**; $2^3 = 8$, $3^2 = 9$ **[1]**
6. 2 packs of sausages **[1]**; 3 packs of bread rolls **[1]**

Page 33
1. $9x + 4y$ **[2]**
2. $4y - 8z + 12$ **[2]**
3. $4x^2 + 2x + 4$ **[2]**
4. 23 **[1]**
5. $8b = 14$ OR $2b = \frac{7}{2}$ **[1]**; $b = \frac{7}{4}$ OR $1\frac{3}{4}$ **[1]**
6. $3p + 6 = 2p + 6$ **[1]**; $p = 0$ **[1]**
7. $\frac{5}{2}x - \frac{2}{3}x = \frac{1}{2} + \frac{1}{3}$ **[1]**; $\frac{15}{6}x - \frac{4}{6}x = \frac{3}{6} + \frac{2}{6}$ **[1]**;
   $x = \frac{5}{11}$ **[1]**

> Look for a common denominator.

8. $6x - 30y + 36$ **[2]** (1 mark for 2 correct terms)
9. $6p - 4q + 12$ **[2]**

> $- \times - = +$

10. $4xz(y - 1)$ **[2]** OR $4z(yx - x)$ **[1]** OR $4x(yz - z)$ **[1]**
11. $(w^2 + 5w + 4)(w - 4)$ **[1]**;
    $w^3 + 5w^2 + 4w - 4w^2 - 20w - 16$ **[1]**;
    $w^3 + w^2 - 16w - 16$ **[1]**
12. $(x + 1)(x + 2)$ **[2]**

13. $6x - 15 + 4x + 12 - 4x$ **[1]**; $6x - 3$ **[1]**;
    $3(2x - 1)$ **[1]**
14. $4xy + 2y = 1 - 3x$ **[1]**;
    $y(4x + 2) = 1 - 3x$ **[1]**;
    $y = \frac{1 - 3x}{4x + 2}$ **[1]**
15. a) $2A = (a + b)h$ **[1]**; $h = \frac{2A}{a + b}$ **[1]**
    b) $h = \frac{48}{5 + 7}$ **[1]**; 4cm **[1]**
16. $(x + 3)(x + 2)$ **[2]**
17. a) $V = \sqrt{50}$ **[1]**; $= 5\sqrt{2}$ OR 7.07 **[1]**
    b) $V^2 = u^2 - 10p$ **[1]**; $u = \sqrt{V^2 + 10p}$ **[1]**

Page 34
1. 1 : 2000 **[1]**
2. $180° \div 9 = 20°$ **[1]**; largest angle = 80° **[1]**
3. 4 people = 6 celery sticks, 1 person =
   $6 \div 4 = 1.5$ sticks (OR 3.5 × 6) **[1]**;
   14 people = 14 × 1.5 = 21 sticks **[1]**
4. a) 1 person takes 24 days (inverse
   proportion) **[1]**; 8 people take
   $24 \div 8 = 3$ days **[1]**
   b) $24 \div 2 = 12$ people **[1]**

1. Density = mass ÷ volume **[1]**;
   Density = 4560 ÷ 400 = 11.4g/cm³ **[1]**
2. Speed = distance ÷ time **[1]**;
   Speed = 200 ÷ 22 = 9.09m/s **[1]**
3. Speed = distance ÷ time **[1]**;
   Speed = 15 ÷ 3 = 5km/h **[1]**
4. Final amount =
   Original amount × $\left(1 + \frac{\text{Rate}}{100}\right)^{\text{Time}}$ **[1]**;
   Final Amount = $4000 × \left(1 + \frac{4}{100}\right)^4$
   = 4000 × 1.04⁴ = £4679.43 **[1]**;
   Compound Interest = £4679.43 − £4000
   = £679.43 **[1]**

Page 35
1.
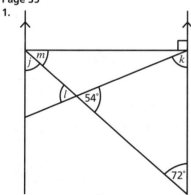
(Marks will not be awarded if reason is
incorrect.) j = 72° (alternate angle) **[1]**;
k = 54° (sum of the interior angles of
a triangle = 180°) **[1]**; l = 54°(vertically
opposite angles are equal) **[1]**; m = 18°
(90° − 72°) **[1]**
2. a) Angle DCB = Angle DAB = 110°
   (opposite angles in a parallelogram
   are equal) **[1]**
   b) Angle ABC = 70° (allied angle to
   Angle BAD) **[1]**
3. $9x = 360°$, $x = 40°$ **[1]**;
   largest angle = 120° **[1]**
4. Exterior angle = $\frac{360}{n} = \frac{360}{10} = 36°$ **[1]**;
   interior angle = 180° − 36° = 144° **[1]**
5. Bearing = 180° + 36° = 216° **[1]**

6. a) Number of sides = 360° ÷ exterior
   angle = 360 ÷ 45 = 8 **[1]**
   b) Octagon **[1]**

Page 36
1. $32 \div 8 = 4$ **[1]**; $32 - 4 = 28$ books **[1]**
2. $\frac{1}{3} + \frac{1}{6} + \frac{1}{4} = \frac{4}{12} + \frac{2}{12} + \frac{3}{12} = \frac{9}{12}$ **[1]**; fraction
   that are horses is $1 - \frac{9}{12} = \frac{3}{12} = \frac{1}{4}$ **[1]**
3. $\frac{1}{6}, \frac{4}{12}, \frac{12}{24}, \frac{2}{3}$ **[2]**
4. a) $\frac{4}{5} × \frac{2}{3} = \frac{8}{15}$m² **[1]**
   b) $\frac{4}{5} + \frac{4}{5} + \frac{2}{3} + \frac{2}{3} = \frac{44}{15}$m **[2]** (1 mark if
   answer is given as a correct
   mixed number)
5. $d = 0.18888...$, $10d = 1.8888...$,
   $100d = 18.8888...$ so $100d - 10d = 17$,
   $90d = 17$ **[1]**; $d = \frac{17}{90}$ **[1]**

1. £24000 − £6000 = £18000 **[1]**;
   $\frac{22}{100} × 18000 = £3960$ **[1]**
2. $\frac{46}{7200} × 100 = 0.639\%$ **[1]**
3. $£4800 × \frac{100}{80}$ **[1]**; = £6000 **[1]**

Page 37
1. a) Fully correct table **[2]** (1 mark for a
   list of 36 outcomes)

	1	2	3	4	5	6
1	1	2	3	4	5	6
2	2	4	6	8	10	12
3	3	6	9	12	15	18
4	4	8	12	16	20	24
5	5	10	15	20	25	30
6	6	12	18	24	30	36

   b) $\frac{13}{36}$ **[1]**
   c) $\frac{8}{36} = \frac{2}{9}$ **[1]**
2. a) $\frac{1}{2}$ **[1]**
   b) $\frac{1}{2}$ **[1]**
3. a) $\frac{305}{500} = \frac{61}{100}$ **[1]**
   b) $\frac{97}{500} × 100 = 19.4$ **[1]**; 19 or 20 **[1]**
4. a)

   b) 0.65 × 0.37 + 0.35 × 0.55 **[2]**;
   = 0.433 **[1]**
5. a)
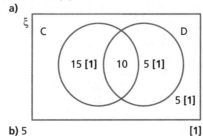
   b) 5 **[1]**

Page 39 Quick Test
1. −4, −8
2. −2, 43
3. 72

Page 41 Quick Test
1. a) $3n + 4$
   b) 154
2. 31, 44
3. $n^3 - 1$

Page 43 Quick Test
1. a) A translation by the vector $\begin{pmatrix} 5 \\ -1 \end{pmatrix}$
   b) A reflection in the line $y = x$
   c) A rotation of 90° anticlockwise
   about (0, 0)

Page 45 Quick Test
1. Draw a line and mark on it two points,
   A and B.

   Open compasses to length AB.

   Put compass point on A and draw an
   arc. Put compass point on B and draw
   an arc.

   Draw a line to join A to the new
   point, C.

   Adjust compasses so less than
   length AB.

   Put compass point on A and draw
   arcs crossing AB and AC at points D
   and E.

   Put compass point on D and draw an
   arc. Put compass point on E and draw
   an arc. Draw a line from A to the new
   point, F.
2.

Page 47 Quick Test
1.

2. Gradient = −2 and y-intercept = 5
3. $y = -1.5x + 14.5$ OR $2y + 3x - 29 = 0$

Answers

Page 49 Quick Test

1.

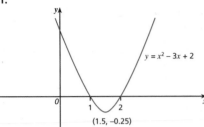

$y = x^2 - 3x + 2$

$(1.5, -0.25)$

2.

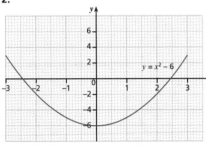

$y = x^2 - 6$

3. a) (5, 2)
b) (8, 3)

Page 51 Quick Test

1. $6x^5$
2. 125
3. $\sqrt{2}$

Page 53 Quick Test

1. Volume = 301.59cm³ (to 2 d.p.) and surface area = 251.33cm² (to 2 d.p.)
2. 24cm²
3. Circumference = 21.99cm (to 2 d.p.) and area = 38.48cm² (to 2 d.p.)

Page 55 Quick Test

1. $\frac{500\pi}{3}$ or 523.6cm³
2. 91.5cm²
3. $\frac{200}{3}$ or 66.7cm³
4. 108π or 339.3cm²

Pages 56–61 Review Questions

Page 56

1. 12 adult tickets = £463.20 and 5 child tickets = £63.80 **[1]**; £527 **[1]**
2. a) 3.119 61 **[1]**
   b) 0.031 196 1 **[1]**
3. −243 **[1]**

1. Numbers are n, $n + 1$, $n + 2$ **[1]**;
   $3n + 3 = 111$, so $n = 36$ **[1]**;
   36, 37, 38 **[1]**
2. a) 2 OR 9 **[1]**
   b) 35 **[1]**
   c) 9 OR 100 **[1]**
   d) 27 **[1]**
   e) 2 OR 13 **[1]**
3. 64 **[1]**
4. $25^2 \times 25^3 = 25^5$ OR 625×15625 $= 9765625 = 25^5$ **[1]**; $m = 5$ **[1]**
5. Peter is right **[1]**; $(2^3)^2 = 2^6 = 64$, $(2^2)^3 = 2^6 = 64$ **[1]**

Page 57

1. $k - g$ **[1]**
2. $12x - 5y$ **[2]**
3. $\frac{7}{9}$ OR 0.7 **[1]**
4. $(2y^2 - y - 21)(6 - y)$ **[1]**;
   $12y^2 - 6y - 126 - 2y^3 + y^2 + 21y$ **[1]**;
   $-2y^3 + 13y^2 + 15y - 126$ **[1]**
5. $b(5a - 3bc)$ **[1]**
6. 5 and 7 are prime numbers, so there are no common factors. **[1]**
7. $x = -6$ **[1]**
8. a) He has only found half of the perimeter. **[1]**
   b) $18x + 14 = 56$ **[1]**; $x = \frac{7}{3} = 2\frac{1}{3}$cm **[1]**
9. a) $\pi \times 2^2 \times 10 = 125.66$cm³ **[1]**
   b) $r^2 = \frac{V}{\pi h}$ **[1]**; $r = \sqrt{\frac{V}{\pi h}}$ **[1]**
   c) $r = \sqrt{\frac{50}{\pi \times 10}}$ **[1]**; $r = 1.26$ (to 3 significant figures) **[1]**

Page 58

1. v^2 is proportional to h so $v^2 = kh$ **[1]**;
   $10^2 = k \times 5$, $100 = 5k$, $k = 20$ **[1]**;
   $v^2 = kh$, $30^2 = 20h$, $900 = 20h$ **[1]**;
   $h = 45$ metres **[1]**
2. 1 part = £60 ÷ 12 = £5 **[1]**; Shares are (5 × £5) = £25, (7 × £5) = £35 **[1]**; Difference = £10 **[1]**
3. 6.2 hours = 6 hours 12 minutes = 372 minutes **[1]**; 93 : 1 **[1]**

1. Type A: Simple Interest =
   $$\frac{\text{Amount} \times \text{Rate} \times \text{Time}}{100} = \frac{400 \times 6 \times 4}{100}$$
   = £96, £400 + £96 = £496 **[1]**; Type B:
   Final Amount =
   Original Amount $\times \left(1 + \frac{\text{Rate}}{100}\right)^{\text{Time}}$ **[1]**;
   $400 \times \left(1 + \frac{5}{100}\right)^4 = 400 \times (1 + 0.05)^4$
   = £486.20 **[1]**; Savings Type A gives the best return after 4 years **[1]**
2. a) Distance = Speed × Time =
   1.5 × 1.5 **[1]**; = 2.25m **[1]**
   b) (11.25 × 8) ÷ 5 = 18km **[1]**;
   $18\text{km/h} = \frac{18000\text{m}}{60 \text{ minutes}}$
   = 300 metres/minute = 5m/s **[1]**;
   After 3 seconds, the mouse is 4.5m through the pipe. After 3 seconds, the cat is 5m through the pipe. Yes, Misty will catch the mouse. **[1]**
3. Food would last (15 × 3 =) 45 days for 1 dog **[1]**; (45 ÷ 5 =) 9 days for 5 dogs **[1]**

Page 59

1. $y + 2y + 3y = 180°$, $6y = 180°$, $y = 30°$ **[1]**; largest angle = 90° **[1]**
2. 80° + 160° + 60° + fourth angle = 360° **[1]**; angle = 60° **[1]**
3. Angles around a point add up to 360°. These angles add up to 364°, so diagram is incorrect. **[1]**
4. Bearing = 180° + 054° = 234° **[1]**
5. Exterior angle = 360° ÷ n **[1]**; 360° ÷ 15 = 24° **[1]**

6. Sum of interior angles (hexagon) = 4 × 180° = 720° **[1]**; $24h = 720°$, $h = 30°$ **[1]**; smallest angle $(2h)$ = 2 × 30° = 60° **[1]**
7. a) True **[1]**
   b) True **[1]**
   c) False **[1]**

Page 60

1. $\frac{3}{5} \times \frac{2}{5} = \frac{6}{25}$ **[1]**; $\frac{4}{25} + \frac{6}{25} = \frac{10}{25} = \frac{2}{5}$ **[1]**

> Remember to use BIDMAS.

2. 36 nails **[1]**
3. $\frac{1}{16}$ **[1]**
4. $\frac{1}{8} \times 344 = 43$ **[1]**; 344 − 43 = 301 biscuits **[1]**

1. $\frac{35}{100} \times £56 = £19.60$ **[1]**; £56 − £19.60 = £36.40 **[1]**
2. Richard is correct **[1]**; $\frac{30}{100} \times £40 = £12$, $\frac{40}{100} \times £30 = £12$ **[1]**
3. $\frac{85000}{215000} \times 100 = 39.5\%$ **[1]**
4. $\frac{80}{65} \times 100$ **[1]**; = £123 **[1]**

Page 61

1. a) 0.3 **[1]**
   b) 50 × 0.3 **[1]**; = 15 **[1]**
   c) No **[1]**; P(Georgia wins) = 0.6, so greater than 0.5, which would be fair. **[1]**
2. $3x - 2 + x + 10x + 4 + 2x = 82$ **[1]**; $16x = 80$, $x = 5$ **[1]**; $2x = 10$ customers **[1]**
3. a) Venn diagram with no crossover **[1]**; All labels correct **[1]**

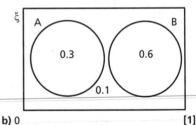

   b) 0 **[1]**
   c) 0.3 + 0.6 **[1]**; = 0.9 **[1]**

> Remember, mutually exclusive means P(A and B) = 0

4. P(vanilla and banana) = $\frac{4}{10} \times \frac{6}{9} = \frac{24}{90} = \frac{4}{15}$ **[1]**;
   P(banana and vanilla) = $\frac{6}{10} \times \frac{4}{9} = \frac{4}{15}$ **[1]**;
   $\frac{4}{15} + \frac{4}{15}$ **[1]**; = $\frac{8}{15}$ **[1]**

Pages 62–65 Practice Questions

Page 62

1. $24 - 4n$ **[2]**
2. −2, 1, 4, 7, 10 **[2]** (1 mark for any three correct)
3. a) $3n + 11$ **[2]**
   b) 311 **[1]**
4. a) $P = 10 \times 2^0$ **[1]**; $P = 10$ **[1]**
   b) $P = 10 \times 2^5$ **[1]**; $P = 320$ **[1]**

c) $2^t = 100$ **[1]**; $t = 7$ hours **[1]**
d) No, as the population will have
   a limit. **[1]**

5. a) 30, 48 **[2]**
   b) 3 **[1]**

6. a)

[1]   **[1]**

   b) $2n + 1$ **[2]**
   c) 201 **[1]**

7.

n	0	1	2	3	4	5
c	–1	4	15	32	55	84
$a + b$	5	11	17	23	29	
$2a$		6	6	6	6	

A complete table or equivalent
working **[1]**; $a = 3$, $b = 2$, $c = –1$ **[1]**;
$3n^2 + 2n – 1$ **[1]**

Page 63

1.

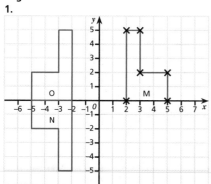

   a) Shape M plotted correctly **[1]**
   b) Shape N plotted correctly **[1]**
   c) Shape O plotted correctly **[1]**
   d) Reflection **[1]**; in the y-axis OR
      mirror line $x = 0$ **[1]**

2. a) Rectangle T is 9cm × 15cm **[1]**;
      Area = 135cm^2 **[1]**
   b) Area R = 15cm^2, Area T = 135cm^2 **[1]**;
      T is 9 times bigger. **[1]**

3. Draw a line and construct the
   perpendicular bisector of the line. **[1]**;
   Bisect the right angle. **[1]**

4. a) A circle **[1]**
   b) An arc of a circle **[1]**
   c) A circle **[1]**
   d) An arc of a circle **[1]**

5. a) Front elevation

[1]

   b) Plan view

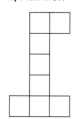

[1]

Page 64

1. Fully correct graph with y-intercept
   at (0, –2) **[1]**; and a straight line
   crossing through points (–4, –18)
   and (4, 14) **[1]**

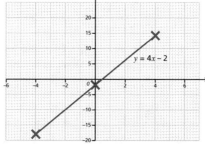

2. Fully correct graph with y-intercept at
   (0, 3) **[1]**; and a straight line crossing
   through points (–2,–7) and (2,13) **[1]**

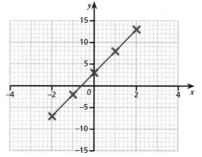

3. Gradient = –2 and y-intercept = (0, 5) **[1]**
4. Fully correct table **[1]**; and accurately
   plotted graph **[1]**

x	–3	–2	–1	0	1	2	3
y	34	17	6	1	2	9	22

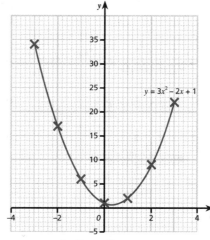

5. $m = \dfrac{-1 - 5}{3 - (-2)} = \dfrac{-6}{5}$ **[1]**;

   $5 = \dfrac{-6}{5} \times \dfrac{-2}{1} + c$, $c = \dfrac{13}{5}$ **[1]**;

   $y = \dfrac{-6}{5}x + \dfrac{13}{5}$ OR $5y = -6x + 13$ **[1]**

6. $m = -1$ **[1]**; $c = 5$ **[1]**; $y = 5 - x$
   OR $x + y = 5$ **[1]**

7. a) $m = \dfrac{R - (R - 5)}{H - (H + 2)}$ **[1]** $= \dfrac{5}{-2}$ **[1]**

   b) $c = R + \dfrac{5}{2}H$ **[1]**

8. a) Minimum (–2, –8) **[1]**;
      maximum (–4, –4) **[1]**
   b) Minimum (0, –8) **[1]**;
      maximum (2, –4) **[1]**
   c) Minimum (0, –4) **[1]**;
      maximum (–2, 0) **[1]**
   d) Minimum (5, –8) **[1]**;
      maximum (3, –4) **[1]**

Page 65

1. a) 5^5 **[1]**
   b) 5^4 **[1]**
   c) $\dfrac{5^6}{5^3}$ **[1]**; $= 5^3$ **[1]**

2. $\dfrac{1}{4}$ (1 mark if 4 seen) **[2]**

3. x^8 **[1]**

4. $4 + 6\sqrt{3} + 6$ **[2]**; $= 10 + 6\sqrt{3}$ **[1]**

5. $27r^6p^3$ **[2]**

1. Area of square = 36cm^2 **[1]**; area of
   circles = $2 \times \pi \times 1.5^2 = 14.137...$ **[1]**;
   shaded region = 36 – 14.137... **[1]**;
   = 21.9cm^2 (to 3 significant figures) **[1]**

2. a) Volume of large cylinder =
      $\dfrac{3}{4} \times 6000\pi = 4500\pi$ **[1]**;
      $4500\pi = \pi \times 15^2 \times h$ **[1]**;
      $h = 20$cm **[1]**
   b) Volume of small cylinder =
      $\dfrac{1}{4} \times 6000\pi = 1500\pi$ **[1]**;
      $1500\pi = \pi \times r^2 \times h = \pi \times r^3$ **[1]**;
      $r = 11.4$cm (to 3 significant figures) **[1]**

3. Volume of cone = $\dfrac{1}{3} \times \pi \times 3^2 \times 7 = 21\pi$ **[1]**;
   volume of the hemisphere =
   $\dfrac{1}{2} \times \dfrac{4}{3} \times \pi \times 3^3 = 18\pi$ **[1]**; volume of
   plastic needed = $21\pi + 18\pi = 39\pi$cm^3 **[1]**

Pages 66–83 Revise Questions

Page 67 Quick Test
1. $y = -2x + 15$
2. $y = 0.5x - 1.5$ OR $2y = x - 3$

Page 69 Quick Test

1.

Answers

2.

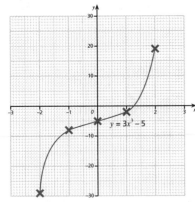

3. a) 75 miles
b) 30 minutes

Page 71 Quick Test

1. $y = -\frac{2}{5}x + \frac{29}{5}$ OR $5y = 29 - 2x$

2. 12 (approximately)

Page 73 Quick Test

1.

$x < 7$

2. $x < 3$

3. Solutions are $x \geqslant 2$ and $x \leqslant -7$

Page 75 Quick Test

1. a) Angle ABC = Angle ADE (corresponding angles are equal); Angle ACB = Angle AED (corresponding angles are equal); Angle DAE is common to both triangles; so triangles are similar (three matching angles).

b) $\frac{5}{10} = \frac{BC}{8}$ so $BC = 4$cm

Page 77 Quick Test

1. 42°
2. Yes, because $8^2 + 15^2 = 17^2$
3. 6.71cm

Page 79 Quick Test

1. a) $\frac{4}{\sin 44} = \frac{AB}{\sin 64}$; $AB = 5.175$cm

b) Angle $B = 72°$, so area = 9.844cm²

Page 81 Quick Test

1. a) 7.76
b) 8
c) 9
d) 3

2. Advantage: cheaper / quicker; disadvantage: could be bias / results not based on whole population

Page 83 Quick Test

1.

(frequency densities: 4.5; 3.4; 2.9)

---

2. a) 11cm
b) 2.5cm

Pages 84–87 Review Questions

Page 84

1. LCM of 3 and 4 = 12 **[1]**; therefore, 29 will be common to both **[1]**

2. a)

Pattern Number	Perimeter (cm)
1	5
2	8
3	11
4	14
60	182 **[1]**
n	$3n + 2$ **[1]**

b) $3n + 2 < 1500$ **[1]**; 499 pentagons **[1]**

3. Yes, Jenny is correct. For n^2, even multiplied by even is even and odd multiplied by odd is odd. **[1]**; For $n^2 + 6$, even + 6 = even and odd + 6 = odd **[1]**

4. $2^n + 2$ **[1]**

Page 85
1. a) $X = (-5, 1)$ **[1]**; $Y = (-3, 5)$ **[1]**; $Z = (-1, 2)$ **[1]**
b) $X = (1, -5)$ **[1]**; $Y = (5, -3)$ **[1]**; $Z = (2, -1)$ **[1]**

2. Lengths of rectangle D: $(3 \times 3 =)$ 9cm and $(5.5 \times 3 =)$ 16.5cm **[1]**; area of rectangle C = 16.5cm², area of rectangle D = 148.5cm² **[1]**; ratio is 1 : 9 **[1]**

3. a) A vertical line **[1]**
b) An arc of a circle **[1]**
c) A horizontal straight line **[1]**
d) An arc of a circle **[1]**

4. A square 4cm × 4cm **[1]**

5. Diameter of lid = 7cm **[1]**; diagram of a rectangle 7cm × 8cm **[1]**

6. a) **[2]**

b) **[2]**

c) **[2]**

Page 86
1. $m = \frac{5 - 8}{\frac{5}{6} - \frac{2}{3}} = -18$ **[1]**;

$8 = -18 \times \left(\frac{2}{3}\right) + c$, $c = 20$ **[1]**;

$y = -18x + 20$ **[1]**

2. $m = \frac{4}{2} = 2$ **[1]**;

y-intercept = $(0, 3)$ **[1]**;

$y = 2x + 3$ **[1]**

---

3. $(x + 1)(x + 3)$ **[1]**; $x = -1$ and $x = -3$ **[1]**;

[1]

4. $m = \frac{3}{4}$ **[1]**; $\left(0, \frac{1}{4}\right)$ **[1]**

5. a) $y = (x + 7)(x - 1)$ **[1]**; $x^2 + 6x - 7$ **[1]**; $a = 6$, $b = -7$ **[1]**
b) $y = (x + 3)^2 - 16$ **[1]**; $(-3, -16)$ **[1]**

6. a)

[1]

b)

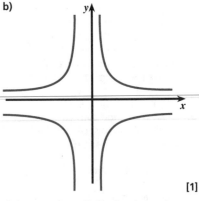

[1]

c) Any two from: Reflection in y-axis; Reflection in x-axis; Rotation 90° clockwise about (0, 0); Rotation 90° anticlockwise about (0, 0). **[2]**

Page 87
1. $\frac{1}{8}a^{-3}b^{15}$ **[1]**; $9a^{-4}b^6$ **[1]**; $\frac{9}{8}a^{-7}b^{21}$ **[1]**

2. $\frac{2}{\sqrt{3}} - \frac{6}{3\sqrt{3}} + 6 \times 4\sqrt{3}$ **[1]**;

$\frac{2}{\sqrt{3}} - \frac{2}{\sqrt{3}} + 24\sqrt{3}$ **[1]**; $24\sqrt{3}$ **[1]**

3. True **[1]**; $x^{-6} = \frac{1}{x^6}$ **[1]**

4. $\frac{1}{5}$ OR 0.2 **[1]**

5. $a = 13$ **[1]**; $b = 14$ **[1]**

1. a) $\frac{1}{2} \times 6 \times 8 \times 9$ **[1]**; 216cm² **[1]**

   b) $\sqrt[3]{216}$ **[1]**; 6×12 **[1]**; = 72cm **[1]**

2. $\pi r^2 = 2\pi r$ **[1]**; $r = 2$ **[1]**

3. $\frac{1}{2}(2x + x) \times 3x \times 20 = 900$ **[1]**;

   $9x^2 = 90$ **[1]**; $x^2 = 10$ **[1]**;

   $x = \sqrt{10}$ or 3.16cm **[1]**

4. $75 = 4 \times \pi \times r^2$ **[1]**;

   $r^2 = \frac{75}{4\pi}$ **[1]**; $r = 2.44$cm (to 3 significant figures) **[1]**

Pages 88–93 Practice Questions

Page 88–89

1. $m = 3$ **[1]**; $5 = 3 + c$, $c = 2$ **[1]**; $y = 3x + 2$ **[1]**

2. Table correctly complete (deduct 1 mark for one error and 2 marks for two errors; 0 marks for more than two errors) **[3]**

Gradient of Line	Gradient of Parallel Line	Gradient of Perpendicular Line
5	5	$-\frac{1}{5}$
3	3	$-\frac{1}{3}$
4	4	$-\frac{1}{4}$
$\frac{8}{9}$	$\frac{8}{9}$	$-\frac{9}{8}$

3. a) Esmai **[1]**

   b) 6mph **[2]** (1 mark for 2 miles in 20 minutes)

   c) 15–20 minutes **[1]**; steepest part of the line **[1]**

   d) 12.5 minutes **[1]**

   e) Esmai and Naval are level for first 5 minutes **[1]**; Esmai then speeds up and overtakes Naval **[1]**; Esmai wins, Naval is second and Gemma is third. **[1]**

4. a) Sketch showing the two curved lines **[1]** intersecting the y-axis at (0, 1) **[1]**

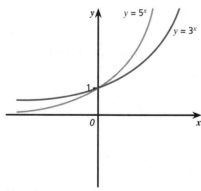

   b) Both pass through point (0, 1) **[1]**

5. a) $16 \div 10$ **[1]**; = 1.6 **[1]**

   b) £1 = \$1.6 **[1]**

6. a) Gradient of radius = $-\frac{5}{4}$ **[1]**;

   Gradient of tangent = $\frac{4}{5}$ **[1]**;

   $-5 = \frac{4}{5} \times 4 + c$, $c = -\frac{41}{5}$ **[1]**;

   $y = \frac{4}{5}x - \frac{41}{5}$ OR $5y = 4x - 41$ **[1]**

   b) $y = \frac{4}{5}x +$ any value **[1]**

7. Area of two triangles, plus a rectangle

   $= \frac{1}{24} + \frac{5}{12} + \frac{5}{24} = \frac{16}{24}$ mile **[1]**;

   Area of trapezium = $\frac{1}{3} = \frac{8}{24}$ mile **[1]**;

   distance travelled = 1 mile **[1]**

> Remember to change the unit of time to hours.

Page 90

1. $-4x > 20$ **[1]**; $x < -5$

   $x < -5$

   –9   –8   –7   –6   –5   –4   **[1]**

> If x is negative the inequality sign will change direction.

2. $x + y \geqslant 5$ **[1]**

3. $0 \leqslant p \leqslant 52$ where p is the number of passengers **[2]**

1. Angle ACB = Angle BDC = 90° **[1]**; Angle ABC = Angle DBC (the angle is common to both triangles) **[1]**; Angle BAC = Angle BCD (180° – Angle B – 90°), so the triangles are similar (three matching angles) **[1]**

2. 64 : 1 **[1]**

Page 91

1. $7.6^2 + 6.7^2 = y^2$ **[1]**; $57.76 + 44.89 = y^2$ **[1]**; $y = \sqrt{(102.65)} = 10.13$km **[1]**

2. $4^2 - 2^2 = f^2$ **[1]**; $16 - 4 = f^2$ **[1]**; $\sqrt{12} = f$, fence height = 3.46m **[1]**; No, because the fence height is 3.46m and Fang can only jump 3m. **[1]**

3. a) AC is 6cm (Pythagorean triple) **[1]**; $6^2 + 6^2 = AD^2$ **[1]**; $\sqrt{72} = AD = 8.49$cm (to 3 significant figures) **[1]**

   b) Scalene triangle **[1]**

4. $10^2 + 6^2 = b^2$ **[1]**; $100 + 36 = b^2$ **[1]**; $b = \sqrt{136} = 11.7$m (to 3 significant figures) **[1]**

5. $1.5^2 + 2^2 = 6.25$ **[1]**; $\sqrt{6.25} = 2.5$ **[1]**; It is a right-angled triangle because Pythagoras' Theorem applies **[1]**

6. $3^2 + 3^2 = d^2$ **[1]**; $d = \sqrt{18} = 4.24$cm **[1]**

7. $\sin 30° = \frac{x}{6}$ (where x is half the length of the hexagonal side) **[1]**; $x = 6 \times \sin 30° = 3$cm, so length of side = 6cm **[1]**

8. $\tan\theta = \frac{5}{3} = 1.6667$ **[1]**; $\theta = 59°$ **[1]**

9. a) Distance (x) of P from base of tower:

   $\tan 35° = \frac{14}{x}$, $x = \frac{14}{\tan 35°}$ **[1]**;

   $x = 19.9941$m **[1]**

   b) Distance (y) of Q from base of tower:

   $\tan 18° = \frac{14}{y}$ **[1]**; $y = 43.0876$m **[1]**

   c) $19.9941^2 + 43.0876^2 = PQ^2$, $399.764 + 1856.541 = PQ^2$ **[1]**; $PQ = 47.5$m **[1]**

Page 92

1. Area = $\frac{1}{2}bc \sin A$,

   $20 = \frac{1}{2} \times (5.7 \times 7.5 \times \sin A)$ **[1]**; $\sin A = \frac{40}{42.75} = 0.9357$ **[1]**; Angle CAB = 69.34° **[1]**; Angle CAB = 69° (to the nearest degree) **[1]**

2. $\frac{14}{\sin 50°} = \frac{12}{\sin A}$ **[1]**;

   $\sin A = \left(\frac{12 \times \sin 50°}{14}\right)$ **[1]**;

   Angle BAC = 41.04° **[1]**

3. $a^2 = b^2 + c^2 - 2bc \cos A$ **[1]**; $a^2 = 80^2 + 30^2 - 2 \times 80 \times 30 \times \cos 30°$ **[1]**; $a^2 = 7300 - 4156.9$ **[1]**; $a = \sqrt{3143.1} = 56.06$m **[1]**; height of the Leaning Tower = 56m **[1]**

4. $\cos A = \frac{(b^2 + c^2 - a^2)}{2bc}$

   $= \frac{(220^2 + 250^2 - 78^2)}{2 \times 220 \times 250}$

   $= \frac{(104\,816)}{110\,000} = 0.95287...$ **[1]**;

   $A = 17.66°$ **[1]**

Page 93

1. a) 50 **[1]**

   b) Correctly drawn histogram, based upon frequency densities of 6, 21, 6.5 and 2.5 to give correct heights **[1]**; correct widths **[1]**; and correct labelling of axes. **[1]**

   c) 21 + 13 + 5 **[1]**; = 39 **[1]**

2. Median = 42 – 44 **[1]**; Q1 = 32 – 34 and Q3 = 52 – 54 **[1]**; correct scale **[1]**; and accurately drawn boxes **[1]**

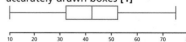

Answers

3. a) Drawn and labelled axes **[1]**; and accurately plotted points. **[1]**

b) Negative correlation – as sunshine increases, rain decreases **[1]**

Pages 94–105 Revise Questions

Page 95 Quick Test
1. $1^2 \times 400 = 400$ OR $(0.9)^2 \times 400 = 324$
2. Width: $2.805 \leq w < 2.815$ cm
   Length: $3.865 \leq l < 3.875$ cm

Page 97 Quick Test
1. $x = -1.4$ or 3.4
2. $x = 1.85$ or -0.18
3. $x = 2 \pm \sqrt{2}$
4. $x = 5.19$ or -0.19

Page 99 Quick Test
1. $x = 2$, $y = 1$
2. $x = 4$, $y = 11$ and $x = -1$, $y = -4$
3. $x = 2$, $y = 2$ and $x = -1$, $y = -1$
4. $18x^2 + 12x + 7$

Page 101 Quick Test
1. $n + (n + 1) = 2n + 1$, $2n$ is even, therefore $2n + 1$ is odd.
2. n^2 and $(n + 1)^2$, n^2 and $n^2 + 2n + 1$. The difference is $2n + 1$, which is an odd number.
3. Set of integers is $n - 2$, $n - 1$, n, $n + 1$, $n + 2$. Mean $= \dfrac{5n}{5} = n$
4. $\dfrac{2}{x} + \dfrac{3}{x^2} = \dfrac{2x}{x^2} + \dfrac{3}{x^2} = \dfrac{2x + 3}{x^2}$

Page 103 Quick Test
1. **a)** $32°$
   b) $56°$
   c) $88°$
   d) $65°$

Page 105 Quick Test
1. **a)** N

b) N

c) N

d) N

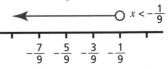

e) N

Pages 106–111 Review Questions

Page 106–107
1. $\dfrac{1}{2} \times 2 \times 30 + \dfrac{1}{2} \times 1 \times 60$ **[1]**;
   $+ \dfrac{1}{2}(30 + 60) \times 1$ **[1]**; $30 + 30 + 45$
   $= 105$ miles **[1]**

> Break the area under the curve down into two triangles and a trapezium.

2. $m_1 = \dfrac{3}{5}$, $m_2 = -\dfrac{5}{3}$ **[1]**; $\dfrac{3}{5} \times -\dfrac{5}{3} = -1$ **[1]**
3. **a) i)** A straight horizontal line **[1]**
   ii) 30 minutes **[1]**
   b) 4.30pm **[1]**
   c) i) $\dfrac{18}{0.5}$ **[1]**; 36km/h **[1]**
   ii) 6pm **[1]**
4. 4 (approx.) **[1]**
5. **a)** $r^2 = 3^2 + 5^2$ **[1]**; $r^2 = 34$ **[1]**;
   $x^2 + y^2 = 34$ **[1]**
   b) Gradient of radius $= \dfrac{5}{3}$ **[1]**; Gradient of tangent $= -\dfrac{3}{5}$ **[1]**; $5y + 3x = 34$ or equivalent **[1]**
6. **a)**

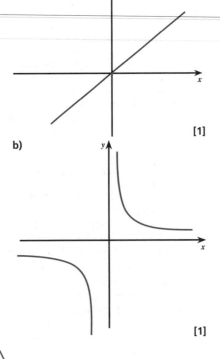

[1]

b)

[1]

7. a) $(x - 3)(x + 3) = x^2 - 9$ **[1]**;
   $(x^2 - 9)(x - 2)$ **[1]**; $x^3 - 2x^2 - 9x + 18$ **[1]**
   b)

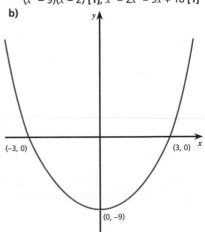

Correct shape graph **[1]**; correct points of intersection **[1]**

8. $m_1 = -\dfrac{1}{2}$, $m_2 = 2$ **[1]**; $6 = 2 \times 3 + c$, $c = 0$ **[1]**; $y = 2x$ **[1]**

Page 108
1. $n = 3, 4, 5, 6, 7$ **[1]**
2. $2x > 8$ **[1]**; $x > 4$ **[1]**
3. $4 \leq y \leq 12$ **[1]**; 4, 5, 6, 7, 8, 9, 10, 11, 12 **[1]**
4. $(x + 8)(x - 3) \geq 0$ **[1]**; one solution is $x \leq -8$ **[1]**; one solution is $x \geq 3$ **[1]**
5. $-9x > 1$, $x < -\dfrac{1}{9}$ **[1]**

   [number line diagram] $x < -\dfrac{1}{9}$

   $-\dfrac{7}{9} \quad -\dfrac{5}{9} \quad -\dfrac{3}{9} \quad -\dfrac{1}{9}$ **[1]**
6. If t is number of televisions, $t > 6$ and $t \leq 20$ **[1]**; $6 < t \leq 20$ **[1]**

1. $56°$ **[1]**; $64°$ **[1]**; $60°$ **[1]**
2. **a)** Isosceles **[1]**
   b) Prove using SSS OR SAS **[3]**
   c) $AD^2 = 10^2 - 6^2$ **[1]**; $AD = 8$cm **[1]**;
   Area $= \dfrac{1}{2}(12 \times 8) = 48$cm^2 **[1]**

Page 109
1. Diagonal length of base:
   $200^2 + 200^2 = d^2$ **[1]**; $d = \sqrt{80000} = 282.84$m; $BC = 141.42$m **[1]**;
   Using triangle TCB: $310^2 - 141.42^2 = TC^2$ **[1]**; $TC = \sqrt{76\,100.38}$, $TC = 276$m **[1]**
2. 5cm **[1]**
3. Third side of the triangle = 5cm (Pythagorean Triple) **[1]**;
   Area $= \dfrac{1}{2}$ base × height, Area $= 6 \times 5$ **[1]**;
   Area $= 30$cm^2 **[1]**
4. **a)** $8^2 + 5^2 = CE^2$ **[1]**; $CE = 9.434$cm **[1]**;
   $\cos \theta = \dfrac{4.717}{13} = 0.36285$ **[1]**; $\theta =$ Angle $PCE = 68.7°$ **[1]**
   b) $13^2 - 4.717^2 = h^2$ **[1]**; Height of P above base = 12.1cm **[1]**
5. $\sin 78° = \dfrac{x}{25}$ **[1]**; $x = 25 \times \sin 78°$ $= 24.45$m **[1]**

Page 110

1. a) Angle C is the smallest **[1]**, because it is opposite the shortest side. **[1]**

   b) $\cos C = \dfrac{(a^2 + b^2 - c^2)}{2ab}$ **[1]**;

   $\cos C = \dfrac{(13^2 + 11^2 - 10^2)}{2 \times 13 \times 11}$ **[1]**; $\cos C$

   $= 0.6643$ **[1]**; Angle $C = 48°$ (to the nearest degree) **[1]**

   c) Area $= \frac{1}{2}ab \sin C$ **[1]**;

   Area $= \frac{1}{2} \times 13 \times 11 \times \sin 48°$ **[1]**;

   Area $= 53.13\text{cm}^2$ **[1]**

2. $b^2 = a^2 + c^2 - 2ac \cos B$ **[1]**;

   $b^2 = 144 + 196 - 2 \times 12 \times 14 \times \cos 50°$ **[1]**;

   $= 340 - 215.976\,6369$

   $= 124.023\,3631$ **[1]**; $b = 11.1\text{km}$ **[1]** OR

   $\dfrac{14}{\sin 75°} = \dfrac{b}{\sin 50°}$ **[1]**; $b = \dfrac{(14 \times \sin 50°)}{\sin 75°}$ **[1]**;

   $= \dfrac{10.724\,6222}{0.965\,925\,826}$ **[1]**; $= 11.1\text{km}$ **[1]**

Page 111

1. a) All frequency values in table correct. Calculated using Frequency Density × Class Width **[2]**

Price (pounds)	Frequency
$0 < P \leqslant 5$	50
$5 < P \leqslant 10$	90
$10 < P \leqslant 20$	70
$20 < P \leqslant 40$	60

   b) $2.5 \times 50 + 7.5 \times 90 + 15 \times 70 + 30 \times 60$

   [1]; $\dfrac{3650}{270}$ **[1]**; £13.52 **[1]**

   c) $\dfrac{270 + 1}{2} = 135.5$ **[1]**; $5 < P \leqslant 10$ **[1]**

2. a) It is biased because she only speaks to library users. **[1]**

   b) A random **[1]** selection of 100 people who live in Malmesbury **[1]**

3. a) Line of best fit drawn as per diagram below **[1]**; Accept 120 to 135 **[1]**

   b) The estimate is not reliable as it is outside the data range. **[1]**

Pages 112–115 Practice Questions

Page 112

1. $\dfrac{500 \times 8}{0.2}$ **[1]**; $= 20\,000$ **[1]**

2. 5kg of carrots cost £4.45, so 1kg of carrots = 89p **[1]**; £3.56 + 5kg of potatoes = £6.11, so 1kg of potatoes = 51p **[1]**; 4kg of potatoes = £2.04 **[1]**

3. £657 = ($h \times 35$) − 218 **[1]**; $\dfrac{657 + 218}{35}$ **[1]**; = 25 hours **[1]**

4. $\sqrt{(6.2^2 - 3.6)} = 5.902\,541\,825$,

   $2.6 \times 0.15 = 0.39$ **[1]**;

   $5.902\,541\,825 \div 0.39 = 15.134\,722\,63$ **[1]**

   a) 15.13 **[1]**

   b) 15.1 **[1]**

5. $\dfrac{£560}{117.5} \times 100$ **[1]**; = £476.60 **[1]**

6. a) $5093 + 3x = 5315$ **[1]**;

   $\dfrac{5315 - 5093}{3} = \dfrac{222 \text{ pence}}{3} = 74 \text{ units}$ **[1]**

   b) $\dfrac{3}{100} \times £50.93 = £1.5279$ **[1]**;

   £50.93 + £1.5279 = £52.46 **[1]**

Page 113

1. $x = \pm 3$ **[1]**

2. a) $(x + 4)^2 - 28$, $p = 4$ **[1]**; $q = -28$ **[1]**

   b) $x = -4 \pm \sqrt{28}$ **[1]**; $x = -4 \pm 2\sqrt{7}$ **[1]**

   c) $(-4, -28)$ **[1]**

3. a) $A = \frac{1}{2}(a + b)h$,

   $14 = \frac{1}{2}(4x - 6 + 2x + 1)\,2x$ **[1]**;

   $14 = \frac{1}{2}(12x^2 - 10x)$ **[1]**;

   $6x^2 - 5x - 14 = 0$ **[1]**

   b) $(6x + 7)(x - 2)$ **[1]**; $x = 2$ **[1]**

If x is a length it must be positive.

1. a) Straight-line graph correct **[1]**; and curved graph correct **[1]**

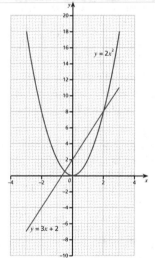

   b) $(-0.5, 0.5)$ **[1]**; $(2, 8)$ **[1]**

When graphs of simultaneous equations are plotted, the x-coordinates of the points of intersection give you your solutions.

2. $y = 1 - 2x$ **[1]**; $1 - 2x = x^2 - 2$ **[1]**;

   $x^2 + 2x - 3 = 0$, $(x + 3)(x - 1) = 0$ **[1]**;

   $(-3, 7)$, $(1, -1)$ OR $x = -3$, $y = 7$ or $x = 1$, $y = -1$ **[1]**

3. a) $f(3x - 2) = (3x - 2)^2 + 1$ **[1]**;

   $fg(x) = 9x^2 - 12x + 5$ **[1]**

   b) $gf(x) = 3(x^2 + 1) - 2$ **[1]**;

   $= 3x^2 + 1 \neq 9x^2 + 12x + 5$ **[1]**

   c) $9x^2 - 12x - 5 = 0$ **[1]**;

   $x = \dfrac{12 \pm \sqrt{(-12)^2 - 4 \times 9 \times (-5)}}{18}$ **[1]**;

   $x = \dfrac{5}{3}$ or $-\dfrac{1}{3}$ **[1]**

Page 114

1. $n^2 + (n + 1)^2 + (n + 2)^2$ **[1]**; $3n^2 + 6n + 5$ **[1]**; $3(n^2 + 2n + 2) - 1$, which is 1 less than a multiple of 3 **[1]**

2. Let n = chosen number **[1]**;

   $\dfrac{3n + 30}{3} = n + 10$ **[1]**; $n + 10 - n = 10$ **[1]**;

3. $\dfrac{3}{x + 2} + \dfrac{9}{(x + 2)(x - 1)}$ **[1]**; $\dfrac{3(x - 1) + 9}{(x + 2)(x - 1)}$ **[1]**;

   $\dfrac{3x - 3 + 9}{(x + 2)(x - 1)} = \dfrac{3(x + 2)}{(x + 2)(x - 1)} = \dfrac{3}{(x - 1)}$ **[1]**

4. $\dfrac{6(3x - 2)}{(3x + 2)(3x - 2)} + \dfrac{4(3x + 2)}{(3x + 2)(3x - 2)}$ **[1]**;

   $\dfrac{18x - 12 + 12x + 8}{9x^2 - 4}$ **[1]**; $\dfrac{30x - 4}{9x^2 - 4}$ **[1]**

Find a common denominator.

5. Area of square $= 36x^2$ **[1]**;

   Area of circle $= \pi(3x)^2 = 9\pi x^2$ **[1]**;

   $36x^2 : 9\pi x^2$ **[1]**; $4 : \pi$ **[1]**

6. $(16n^2 + 8n + 1) - (16n^2 - 8n + 1)$ **[1]**;

   $16n$ **[1]**; $8(2n)$ which is a multiple of 8 **[1]**

7. $(2n + 2)^2 - (2n)^2$ **[1]**; $8n + 4$ **[1]**; $4(2n + 1)$ which is a multiple of 4 **[1]**

8. $a + b = ab$ **[1]**; $b + 2 + b = b(b + 2)$ **[1]**;

   $b^2 = 2$, therefore b is not an integer, and $a = \sqrt{2} + 2$, so a is not an integer. **[1]**

Page 115

1. a) $a = 34°$ ($90° - 56°$) **[1]**

   b) $b = 102°$ ($204° \div 2$) **[1]**; $c = 78°$ ($180° - 102°$) **[1]**

   c) $d = 72°$ (alternate segment) **[1]**; $e = 50°$ ($180° - 72° - 58°$) **[1]**

   d) $f = 37°$ (angle subtended by a chord) **[1]**; $g = 21°$ (angle subtended by a chord) **[1]**

   e) $h = 42°$ ($90° - ((180° - 84°) \div 2)$ $= 90° - 48° = 42°$) **[1]**; $i = 42°$ (isosceles triangle) **[1]**

   f) $j = 36°$ (isosceles triangle so $5j = 180°$) **[1]**

1. a) $a + \frac{1}{2}(c - a)$ **[2]**;

   $\frac{1}{2}a + \frac{1}{2}c = \frac{1}{2}(a + c)$ **[1]**

   b) $\overrightarrow{DE} = \frac{1}{2}a + \frac{1}{2}(c - a)$ **[1]**;

   $= \frac{1}{2}a + \frac{1}{2}c - \frac{1}{2}a = \frac{1}{2}c$ **[1]**;

   $\overrightarrow{AC} = c$, so $\overrightarrow{DE}$ is parallel to $\overrightarrow{AC}$ **[1]**

Answers

2. a) Scalar **[1]**
   b) Vector **[1]**
   c) Vector **[1]**

Page 116

1. $805.975 \div 6.14 = 131.266\,286\,6$ **[1]**
   a) 131.27 **[1]**
   b) 131 **[1]**

2. a) Per day £84 (adults) + £35.70 (child)
      = £119.70 **[1]**; Holiday costs
      9 × £119.70 **[1]**; = £1077.30 **[1]**
   b) 5% of £1077.30 **[1]**; = £53.87 **[1]**
   c) 11.10am + 4 hours 20 minutes + 1
      hour 34 minutes, flight arrives in
      Turkey at 5.04pm (17:04) in English
      time **[1]**; Turkish arrival time =
      7.04pm (19:04) **[1]**

3. 18.0625cm² **[1]**

4. a) Option 1: £146 + £80 × 12 = £1106,
      $\frac{95}{100}$ × £1106 = £1050.70 **[1]**;
      Option 2: £548 + £43 × 12 = £1064
      $\frac{95}{100}$ × £1064 = £1010.80 **[1]**;
      Option 2 is cheaper **[1]**
   b) £1050.70 – £1010.80 = £39.90 **[1]**

Page 117

1. a) $y = 6$ **[1]**
   b) $x^2 + (3)^2 = 36$, $x^2 = 27$ **[1]**; $x = 5.2$ or
      -5.2 **[1]**; (–5.2, 3) and (5.2, 3) **[1]**

2. a) $(x - 6)^2 - 10p = -6$ **[1]**; $q = -10$ **[1]**;
   b) $(x - 6)^2 = 10$, $x - 6 = \pm\sqrt{10}$ **[1]**;
      $x = 6 + \sqrt{10}$, $x = 6 - \sqrt{10}$ **[1]**

3. $\frac{4(x + 2)}{x(x + 2)} + \frac{4x}{x(x + 2)} = 3$ **[1]**; $\frac{8x + 8}{x^2 + 2x} = 3$,
   $8x + 8 = 3x^2 + 6x$ **[1]**;
   $3x^2 - 2x - 8 = 0$ **[1]**; $(3x + 4)(x - 2)$,
   $x = -\frac{4}{3}$ and $x = 2$ **[1]**

4. $\frac{1}{2} \times \frac{x}{5} \times \frac{x}{15} = 4$ **[1]**; $x^2 = 600$ **[1]**;
   $x = \sqrt{600}$, $x = 10\sqrt{6}$ **[1]**

5. $x = \frac{7 \pm \sqrt{(-7)^2 - 4 \times 1 \times -15}}{2}$ **[1]**;
   $x = -1.72$ and 8.72 (to 3 significant
   figures) **[1]**

1. a) Correct curve **[1]**; correct straight
      line **[1]**

   b) $x = 1.6$ to 1.8 **[1]**; (–0.3) to (–0.5) **[1]**

2. $y = x + 4$, $x + 4 = x^2 + 2$ **[1]**;
   $x^2 - x - 2 = 0$ **[1]**; $(x - 2)(x + 1) = 0$,
   $x = 2$, $y = 6$ **[1]**;
   $x = -1$, $y = 3$ **[1]**

Page 118

1. a) $3n + 1$ **[1]**
   b) $(3n + 1)(3(n + 1) + 1)$ **[1]**;
      $9n^2 + 15n + 4$ **[1]**;
      $3(3n^2 + 5n + 1) + 1$ **[1]**

2. $(n + 1)^2 - n^2$ **[1]**; $= n^2 + 2n + 1 - n^2$
   $= 2n + 1$ **[1]**; $= n + (n + 1)$ **[1]**

3. $\frac{(x + 3)(x + 2)}{x^2 + 2x}$ **[1]**; $\frac{(x + 3)(x + 2)}{x(x + 2)}$ **[1]**;
   $\frac{x + 3}{x}$ **[1]**

4. Any odd number –7 **[1]**

5. Area = x^2, diagonal = $\sqrt{x^2 + x^2}$
   = $\sqrt{2x^2}$ **[1]**; $\sqrt{2x^2} \times \sqrt{2x^2} = 2x^2$,
   $\frac{2x^2}{2} = x^2$ **[1]**

6. $9n^2 + 6n + 1 - (9n^2 - 6n + 1)$ **[1]**;
   $12n$ **[1]**; $4(3n)$ **[1]**

7. a) $n^2 + 1$ **[1]**
   b) $(n^2 + 1) + ((n + 1)^2 + 1)$ **[1]**;
      $2n^2 + 2n + 3$ **[1]**; $2(n^2 + n + 1) + 1$ **[1]**

8. $\frac{2}{x - 2} - \frac{8}{(x + 2)(x - 2)}$ **[1]**; $\frac{2(x + 2)}{(x - 2)(x + 2)}$
   $- \frac{8}{(x + 2)(x - 2)}$, $\frac{2x - 4}{(x + 2)(x - 2)}$ **[1]**;
   $\frac{2(x - 2)}{(x + 2)(x - 2)}$, $\frac{2}{(x + 2)}$ **[1]**

9. $(2n + 1)^2$ **[1]** $= 4n^2 + 4n + 1$ **[1]**;
   $4(n^2 + n) + 1$ **[1]**

10. The sum of any two primes that are
    not 2, e.g. 3 + 5 = 8 **[1]**

Page 119

1. a) $k = 52°$ (alternate segment) **[1]**
   b) $l = 90°$ (angle subtended from
      diameter) **[1]**
   c) $m = 48°$ (96° ÷ 2) **[1]**; $n = 42°$
      ((180° – 96°) ÷ 2) **[1]**

2. a) a = radius **[1]**; b = circumference **[1]**;
      c = arc **[1]**; d = segment **[1]**;
      e = sector **[1]**
   b) Arc length = $\frac{\theta}{360} \times 2 \times \pi \times r$ **[1]**;
      Arc length = $\frac{40}{360} \times 2 \times 3.14 \times 10$
      = 6.977 777 778 **[1]**;
      Arc length = 6.98cm **[1]**

1. a) $\overrightarrow{AC} = 4c - 4a$ **[1]**
   b) $\overrightarrow{AP} = \frac{1}{4}(4c - 4a) = c - a$ **[1]**
   c) $\overrightarrow{DP} = 4a + c - a = 3a + c$ **[1]**

2. a) True **[1]**
   b) False **[1]**

1. Trapezium below line C shaded **[1]**
2. $3x^2(x + 2)$ **[1]**
3. $3x + 9 - 5 = 2x - 4$ **[1]**; $3x + 4$
    $= 2x - 4$ **[1]**; $x = -8$ **[1]**
4. $\frac{5}{7}\left(\frac{1}{5} - \frac{2}{5}\right) + \frac{18}{7}$ **[1]**; $\frac{17}{7}$ **[1]**
5. $-5x^2 + 10x - 2y$ **[2]**
6. $4p^2y + 6py$ **[2]**
7. $x - 3 = -\frac{3}{5}$ **[1]**; $x = \frac{12}{5}$ **[1]**
8. $a = 6$, $b = -16$, $c = -14$, $d = 24$,
    $6x^3 - 16x - 14x + 24$ **[2]** (1 mark for any
    two correct terms)
9. a) $(x - 2)(x + 1)$ **[1]**
   b) $(2x - 1)(x - 2)$ **[1]**
   c) $\frac{(2x - 1)(x - 2)}{(x - 2)(x + 1)}$ **[1]**; $\frac{2x - 1}{x + 1}$ **[1]**
10. $30 = 60 - q^2$ **[1]**; $q = \sqrt{30}$ **[1]**

> The question asks for an exact value,
> so leave your answer in surd form.

11. a) $(x + 4)(x - 4)$ **[1]**
    b) $x = 4$ **[1]**; $x = -4$ **[1]**
12. Ethan is correct **[1]**; $r^2 = \frac{25}{\pi}$,
    $r = \sqrt{\frac{25}{\pi}} = \frac{5}{\sqrt{\pi}}$ **[1]**

> You must square root both the
> numerator and denominator.

13. $\left(\sqrt{5}\right)^2 + 3\sqrt{5}$ **[1]**; $5 + 3\sqrt{5}$ **[1]**
14. 6.677, 6.767, 6.77, 6.8 **[1]**
15. a) 5 × 3 × 3 **[1]**; = 5×3^2 **[1]**
    b) 3 × 5 × 7 **[1]**
    c) 3 × 5 **[1]**; = 15 **[1]**
    d) x^2y^2 **[2]**
16. $\frac{31}{6} - \frac{7}{3}$ **[1]**; $\frac{31}{6} - \frac{14}{6}$ **[1]**; $\frac{17}{6}$ **[1]**
17. $P = 1.1 \times 1.1$ **[1]** = 1.21, 21% increase **[1]**
18. Dave's Dongles: £16.20 for six months **[1]**;
    £18 for following six months, total
    £205.20 **[1]**; Ian's Internet: £14.62 for
    four months **[1]**; £17.20 for following
    8 months, total £196.08 **[1]**; Ian's
    Internet is cheaper **[1]**
19. a) 1408.44 **[1]**
    b) 0.140844 **[1]**
20. $1.5 \times 10^8 \div 3.5 \times 10^4$ **[1]**;
    4285.7 (to 1 d.p.) = 4.2857×10^3 **[1]**
21. Speed = $\frac{\text{Distance}}{\text{Time}}$ **[1]**
22. Exterior angle = $\frac{360°}{20°}$ = 18° **[1]**;
    interior angle = 180° – 18° = 162° **[1]**
23. $3n + 5$ **[1]**

24. a) $2 \leqslant a < 4$

b) $(1 \times 256) + (3 \times 219) + (5 \times 165) +$
$(7 \times 120) + (9 \times 65) = 3163$ **[1]**;
$3163 \div 825$ **[1]**; $= 3.83$ (to 2 d.p.) **[1]**

> Remember to use midpoints to
> calculate the mean of grouped data.

c)

Correct widths **[1]**; correct frequency
densities: 128, 109.5, 82.5, 60, 32.5 **[1]**

d) No **[1]**; the median is now the 452nd
value, which is in the same group. **[1]**

25. a) $F \alpha \dfrac{k}{x^2}$, $4 = \dfrac{k}{9}$ **[1]**; $k = 36$,

$F = \dfrac{36}{x^2}$ **[1]**

b) $F = \dfrac{36}{25}$ or 1.44 **[1]**

c) $20 = \dfrac{36}{x^2}$ **[1]**; $x^2 = \dfrac{36}{20}$,

$x = \sqrt{\dfrac{9}{5}}$ or 1.34 **[1]**

26. a) $x^2 = 6^2 + 9^2 - 2 \times 6 \times 9 \times \cos 37°$ **[1]**;
$x^2 = 30.74...$ **[1]**; $x = 5.55$cm
(to 3 significant figures) **[1]**

b) $\dfrac{1}{2} \times 6 \times 9 \times \sin 37°$ **[1]**; $= 16.25$cm² **[1]**

c) 9 **[1]**

27. $x = 0$ **[1]**

28. a) $3 : 8$ **[1]**

b) $\dfrac{20}{74} \times 100$ **[1]**; $= 27\%$
(to the nearest 1%) **[1]**

29. a) 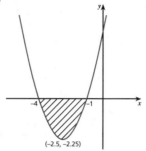 **[1]**

b) $(x + 1)(x + 4) \leqslant 0$ **[1]**;
$-4 \leqslant x \leqslant -1$ **[1]**

c) Shaded region on graph above **[1]**

30. a) i) $2b - 2a$ **[1]**

ii) $2a$ **[2]**

b) $4a - b$ **[2]**

c) $\overrightarrow{DY} = 4a - \dfrac{2}{3}(2b - 2a)$

$= \dfrac{16}{3}a - \dfrac{4}{3}b$ **[1]**;

$\overrightarrow{DX} = 4a - b$ **[1]**;

$DY = \dfrac{4}{3}DX$, therefore,

DXY is a straight line **[1]**

31. $x^2 - 1 = 3x + 3$ **[1]**;
$(x - 4)(x + 1) = 0$ **[1]**;
$x = 4$, $y = 15$ **[1]**;
$x = -1$, $y = 0$ **[1]**

32. a) 33° **[1]**; radius and tangent meet at
90° **[1]**

b) 57° **[1]**; Triangle in semicircle, angle
at circumference = 90° or alternate
segment theorem **[1]**

33. $\dfrac{1}{5}t^2$ **[2]**

34. $2 \times \pi \times 6 = 12\pi$ **[1]**; $12\pi \times \dfrac{115}{360}$ **[1]**;
$12\pi \times \dfrac{115}{360} + 12$ **[1]**; 24.0 (to 3 significant
figures) **[1]**

35. $\dfrac{6}{11} \times \dfrac{5}{11}$ **[1]**; $\dfrac{30}{121} \times 2$ **[1]**; $\dfrac{60}{121}$ **[1]**

> The bead is replaced each time.

36. a) 1.55×2.655 **[1]**; $= 4.11525$ **[1]**

b) $1.45 \div 2.655$ **[1]**; $= 0.546$ (to 3 d.p.) **[1]**

37. a) $(-2, 0)$ **[1]**; $(-1, 0)$ **[1]**

b) $(x + 1.5)^2 - 0.25$ **[2]**; $(-1.5, -0.25)$ **[1]**

c) $(-3.5, -0.25)$ **[1]**

Glossary and Index

Qualitative data   data that is non-numerical, e.g. colour of cars   80

Quantitative data   data that is numerical, e.g. number of cars   80

Quartiles   the three points that divide a range of data into four quarters   82–83

R

Radius / radii (pl.)   the length of a straight line from the centre of a circle to the circumference   45, 52–53, 55, 71, 102–103

Random   when each object / person has an equal chance of being selected   28

Range   the spread of data   80–81

Rate of change   the rate at which one quantity changes in relation to another; the change in the y-value divided by the change in the x-value   67, 70–71

Ratio   the relative amounts of two or more things, shown in the form A : B   16–17, 67, 74, 101

Rational number   a number that can be written exactly in fraction or decimal forms, e.g. $\frac{1}{4}$ = 0.25   25

Rationalise   to make an irrational denominator into a rational number by multiplying both the numerator and denominator by a surd (square root)   51

Reciprocal   the reciprocal of a number (n) is 1 divided by that number, i.e. $\frac{1}{n}$; a reciprocal function takes the form $y = \frac{a}{x}$ 25, 69

Recurring   repeating   25

Recursive   a sequence in which the terms are used in the rule to determine other terms   38

Reflection   a transformation that produces a mirror image of the original object   42

Reflex   an angle between 180° and 360°   20

Region   an area when dealing with graphical inequalities, often bounded by lines   72–73

Regular   shapes in which all the sides are equal and all angles are equal   22

Relative frequency = $\dfrac{\text{frequency of a particular outcome}}{\text{total number of trials}}$ 28

Rhombus   a quadrilateral in which all sides are equal and opposite sides are parallel   21

Right-angled   a triangle with one 90° angle   20, 76–77

Root   a root is the inverse operation to an index or power, e.g. $\sqrt{x}$ is the inverse of x^2 and $\sqrt[3]{x}$ is the inverse of x^3 10, 50–51

Roots   the points of intersection, where the curve of a quadratic graph crosses the x-axis   48–49

Rotation   a transformation that turns an object; every point of the object is turned through the same angle about a given point   42

S

Sample   a proportion of a population   80

Sample space diagram   a probability diagram that contains all the possible outcomes of an experiment   28, 30

Scalar   a numerical quantity   104

Scale factor   the ratio by which an object is made bigger or smaller   43

Scalene   a triangle with no equal sides or equal angles   20

Scatter diagram   a statistical graph that compares two variables by plotting one against the other   81

Secondary data   data used for an investigation that has been collected by a third party   80

Sector   the area between an arc and two radii   103

Segment   the area between an arc and a chord   103

Sequence   a collection of terms that follow a rule or pattern   38–41

Significant figures   a means of approximating numbers; the first significant figure is the first non-zero figure (working from left to right)   94–95

Similar   similar figures are identical in shape but differ in size   43, 54, 74–75

Simplify   to make simpler, usually by cancelling down   16

Simultaneous equations   two or more equations that are true at the same time and must therefore be solved together   98–99

Sine   a trigonometric ratio stating $\sin\theta = \dfrac{\text{opposite side}}{\text{hypotenuse}}$ 77–79

Sine rule   a rule used to calculate unknown sides or angles in a triangle that is not right-angled, i.e. $\dfrac{a}{\sin A} = \dfrac{b}{\sin B} = \dfrac{c}{\sin C}$ 78–79

Speed   a measure of how fast something is moving, i.e. $\dfrac{\text{distance}}{\text{time}}$ 18

Sphere   a 3D shape that is round, like a ball; at every point, its surface is equidistant from its centre   55

Square number   the result of multiplying a number by itself, e.g. $4^2 = 4 \times 4 = 16$ 10, 14, 39, 50, 100

Square root   the inverse of a square number, e.g. $4^2 = 16$ and $\sqrt{16} = 4$ 10, 50–51

Standard form   a number written in the form of $a \times 10^n$ where a is a number between 1 and 10; for large numbers n is positive and for numbers less than 1, n is negative   9

Stratified   a sample taken from each subgroup of a population in the same proportion as they occur in the population   80–81

Subject (of a formula)   the subject of a formula is the variable that appears on its own; formulae can be rearranged to make different variables the subject   15, 79, 98

Subtended   formed from / created by   102

Surd   a number written as a square root; a surd is an exact number   51

T

Tangent   a trigonometric ratio stating $\tan\theta = \dfrac{\text{opposite side}}{\text{adjacent side}}$ 77

Tangent (line)   a straight line touching the circumference of a circle in one place only   70–71, 79, 102–103

Term   in an expression, any of the quantities connected to each other by an addition or subtraction sign; in a sequence, any of the values connected to each other by a pattern or rule   12–14, 38–41, 78, 96–98

Terminating decimal   is a finishing decimal, i.e. with a finite number of digits, e.g. 0.75, 0.36   25

Term-to-term rule   a rule that links a term in a sequence to the previous one   38

Theta (θ)   a Greek symbol used in trigonometry to represent an unknown angle   77

Transformation   an action that brings about a change to the position, size or orientation of a shape, i.e. translation, rotation, reflection and/or enlargement   42–43, 49

Translation   a transformation that moves the object, but does not turn it   42

Trapezium   a quadrilateral with one pair of parallel sides   21, 52

Tree diagram   a diagram that represents probability, with each branch showing a different possible outcome   30–31

Turning point   the point at which the curve of a graph changes direction   48–49

Two-way table   a table that represents two variables in a set of data   29

U

Upper bound   the upper limit of a rounded number   95

V

Variable   a quantity that can have many values; often written as a letter   12, 67, 70, 81

Vector   a quantity that has both magnitude (size) and direction (shown by an arrow)   42, 104–105

Velocity   a compound measure that represents speed and direction of travel   69

Venn diagram   a diagram that uses circles to represent sets, with the position and overlap of the circles showing the relationships between the sets   29

Vertex / vertices (pl)   the corner of a shape   24, 54

Z

Zero term   the theoretical term that would come before the first term of a given sequence   40